GROUPS OF GALAXIES

A SERIES OF BOOKS ON RECENT DEVELOPMENTS IN ASTRONOMY AND ASTROPHYSICS

A.S.P. CONFERENCE SERIES
BOARD OF EDITORS

Dr. Sallie L. Baliunas, Chair
Dr. John P. Huchra
Dr. Roberta M. Humphreys
Dr. Catherine A. Pilachowski

© Copyright 1995 Astronomical Society of the Pacific
390 Ashton Avenue, San Francisco, California 94112

All rights reserved

Printed by BookCrafters, Inc.

First published 1995

Library of Congress Catalog Card Number: 94-73370
ISBN 0-937707-89-9

D. Harold McNamara, Managing Editor of Conference Series
408 ESC Brigham Young University
Provo, UT 84602
801-378-2298

A SERIES OF BOOKS ON RECENT DEVELOPMENTS IN ASTRONOMY AND ASTROPHYSICS

Vol. 1-Progress and Opportunities in Southern Hemisphere Optical Astronomy: The CTIO 25th Anniversary Symposium
ed. V. M. Blanco and M. M. Phillips ISBN 0-937707-18-X

Vol. 2-Proceedings of a Workshop on Optical Surveys for Quasars
ed. P. S. Osmer, A. C. Porter, R. F. Green, and C. B. Foltz ISBN 0-937707-19-8

Vol. 3-Fiber Optics in Astronomy
ed. S. C. Barden ISBN 0-937707-20-1

Vol. 4-The Extragalactic Distance Scale: Proceedings of the ASP 100th Anniversary Symposium
ed. S. van den Bergh and C. J. Pritchet ISBN 0-937707-21-X

Vol. 5-The Minnesota Lectures on Clusters of Galaxies and Large-Scale Structure
ed. J. M. Dickey ISBN 0-937707-22-8

Vol. 6-Synthesis Imaging in Radio Astronomy: A Collection of Lectures from the Third NRAO Synthesis Imaging Summer School
ed. R. A. Perley, F. R. Schwab, and A. H. Bridle ISBN 0-937707-23-6

Vol. 7-Properties of Hot Luminous Stars: Boulder-Munich Workshop
ed. C. D. Garmany ISBN 0-937707-24-4

Vol. 8-CCDs in Astronomy
ed. G. H. Jacoby ISBN 0-937707-25-2

Vol. 9-Cool Stars, Stellar Systems, and the Sun. Sixth Cambridge Workshop
ed. G. Wallerstein ISBN 0-937707-27-9

Vol. 10-The Evolution of the Universe of Galaxies. The Edwin Hubble Centennial Symposium
ed. R. G. Kron ISBN 0-937707-28-7

Vol. 11-Confrontation Between Stellar Pulsation and Evolution
ed. C. Cacciari and G. Clementini ISBN 0-937707-30-9

Vol. 12-The Evolution of the Interstellar Medium
ed. L. Blitz ISBN 0-937707-31-7

Vol. 13-The Formation and Evolution of Star Clusters
ed. K. Janes ISBN 0-937707-32-5

Vol. 14-Astrophysics with Infrared Arrays
ed. R. Elston ISBN 0-937707-33-3

Vol. 15-Large-Scale Structures and Peculiar Motions in the Universe
ed. D. W. Latham and L. A. N. da Costa ISBN 0-937707-34-1

Vol. 16-Atoms, Ions and Molecules: New Results in Spectral Line Astrophysics
ed. A. D. Haschick and P. T. P. Ho ISBN 0-937707-35-X

Vol. 17-Light Pollution, Radio Interference, and Space Debris
ed. D. L. Crawford ISBN 0-937707-36-8

Vol. 18-The Interpretation of Modern Synthesis Observations of Spiral Galaxies
ed. N. Duric and P. C. Crane ISBN 0-937707-37-6

Vol. 19-Radio Interferometry: Theory, Techniques, and Application, IAU Colloquium 131
ed. T. J. Cornwell and R. A. Perley ISBN 0-937707-38-4

Vol. 20-Frontiers of Stellar Evolution, celebrating the 50th Anniversary of McDonald Observatory
ed. D. L. Lambert ISBN 0-937707-39-2

Vol. 21-The Space Distribution of Quasars
ed. D. Crampton
ISBN 0-937707-40-6

Vol. 22-Nonisotropic and Variable Outflows from Stars
ed. L. Drissen, C. Leitherer, and A. Nota
ISBN 0-937707-41-4

Vol. 23-Astronomical CCD Observing and Reduction Techniques
ed. S. B. Howell
ISBN 0-937707-42-4

Vol. 24-Cosmology and Large-Scale Structure in the Universe
ed. R. R. de Carvalho
ISBN 0-937707-43-0

Vol. 25-Astronomical Data Analysis Software and Systems I
ed. D. M. Worrall, C. Biemesderfer, and J. Barnes
ISBN 0-937707-44-9

Vol. 26-Cool Stars, Stellar Systems, and the Sun, Seventh Cambridge Workshop
ed. M. S. Giampapa and J. A. Bookbinder
ISBN 0-937707-45-7

Vol. 27-The Solar Cycle
ed. K. L. Harvey
ISBN 0-937707-46-5

Vol. 28-Automated Telescopes for Photometry and Imaging
ed. S. J. Adelman, R. J. Dukes, Jr., and C. J. Adelman
ISBN 0-937707-47-3

Vol. 29-Workshop on Cataclysmic Variable Stars
ed. N. Vogt
ISBN 0-937707-48-1

Vol. 30-Variable Stars and Galaxies, in honor of M. S. Feast on his retirement
ed. B. Warner
ISBN 0-937707-49-X

Vol. 31-Relationships Between Active Galactic Nuclei and Starburst Galaxies
ed. A. V. Filippenko
ISBN 0-937707-50-3

Vol. 32-Complementary Approaches to Double and Multiple Star Research, IAU Collouquium 135
ed. H. A. McAlister and W. I. Hartkopf
ISBN 0-937707-51-1

Vol. 33-Research Amateur Astronomy
ed. S. J. Edberg
ISBN 0-937707-52-X

Vol. 34-Robotic Telescopes in the 1990s
ed. A. V. Filippenko
ISBN 0-937707-53-8

Vol. 35-Massive Stars: Their Lives in the Interstellar Medium
ed. J. P. Cassinelli and E. B. Churchwell
ISBN 0-937707-54-6

Vol. 36-Planets and Pulsars
ed. J. A. Phillips, S. E. Thorsett, and S. R. Kulkarni
ISBN 0-937707-55-4

Vol. 37-Fiber Optics in Astronomy II
ed. P. M. Gray
ISBN 0-937707-56-2

Vol. 38-New Frontiers in Binary Star Research
ed. K. C. Leung and I. S. Nha
ISBN 0-937707-57-0

Vol. 39-The Minnesota Lectures on the Structure and Dynamics of the Milky Way
ed. Roberta M. Humphreys
ISBN 0-937707-58-9

Vol. 40-Inside the Stars, IAU Colloquium 137
ed. Werner W. Weiss and Annie Baglin
ISBN 0-937707-59-7

Vol. 41-Astronomical Infrared Spectroscopy: Future Observational Directions
ed. Sun Kwok
ISBN 0-937707-60-0

Vol. 42-GONG 1992: Seismic Investigation of the Sun and Stars
ed. Timothy M. Brown
ISBN 0-937707-61-9

Vol. 43-Sky Surveys: Protostars to Protogalaxies
ed. B. T. Soifer   ISBN 0-937707-62-7

Vol. 44-Peculiar Versus Normal Phenomena in A-Type and Related Stars
ed. M. M. Dworetsky, F. Castelli, and R. Faraggiana   ISBN 0-937707-63-5

Vol. 45-Luminous High-Latitude Stars
ed. D. D. Sasselov   ISBN 0-937707-64-3

Vol. 46-The Magnetic and Velocity Fields of Solar Active Regions, IAU Colloquium 141
ed. H. Zirin, G. Ai, and H. Wang   ISBN 0-937707-65-1

Vol. 47-Third Decinnial US-USSR Conference on SETI
ed. G. Seth Shostak   ISBN 0-937707-66-X

Vol. 48-The Globular Cluster-Galaxy Connection
ed. Graeme H. Smith and Jean P. Brodie   ISBN 0-937707-67-8

Vol. 49-Galaxy Evolution: The Milky Way Perspective
ed. Steven R. Majewski   ISBN 0-937707-68-6

Vol. 50-Structure and Dynamics of Globular Clusters
ed. S. G. Djorgovski and G. Meylan   ISBN 0-937707-69-4

Vol. 51-Observational Cosmology
ed. G. Chincarini, A. Iovino, T. Maccacaro, and D. Maccagni   ISBN 0-937707-70-8

Vol. 52-Astronomical Data Analysis Software and Systems II
ed. R. J. Hanisch, J. V. Brissenden, and Jeannette Barnes   ISBN 0-937707-71-6

Vol. 53-Blue Stragglers
ed. Rex A. Saffer   ISBN 0-937707-72-4

Vol. 54-The First Stromlo Symposium: The Physics of Active Galaxies
ed. Geoffrey V. Bicknell, Michael A. Dopita, and Peter J. Quinn   ISBN 0-937707-73-2

Vol. 55-Optical Astronomy from the Earth and Moon
ed. Diane M. Pyper and Ronald J. Angione   ISBN 0-937707-74-0

Vol. 56-Interacting Binary Stars
ed. Allen W. Shafter   ISBN 0-937707-75-9

Vol. 57-Stellar and Circumstellar Astrophysics
ed. George Wallerstein and Alberto Noriega-Crespo   ISBN 0-937707-76-7

Vol. 58-The First Symposium on the Infrared Cirrus and Diffuse Interstellar Clouds
ed. Roc M. Cutri and William B. Latter   ISBN 0-937707-77-5

Vol. 59-Astronomy with Millimeter and Submillimeter Wave Interferometry
ed. M. Ishiguro and Wm. J. Welch   ISBN 0-937707-78-3

Vol. 60-The MK Process at 50 Years: A Powerful Tool for Astrophysical Insight
ed. C. J. Corbally, R. O. Gray, and R. F. Garrison   ISBN 0-937707-79-1

Vol. 61-Astronomical Data Analysis Software and Systems III
ed. Dennis R. Crabtree, R. J. Hanisch, and Jeannette Barnes   ISBN 0-937707-80-5

Vol. 62-The Nature and Evolutionary Status of Herbig Ae / Be Stars
ed. P. S. Thé, M. R. Pérez, and E. P. J. van den Heuvel   ISBN 0-937707-81-3

Vol. 63-Seventy-Five Years of Hirayama Asteroid Families: The role of Collisions in the Solar System History
ed. R. Binzel, Y. Kozai, and T. Hirayama   ISBN 0-937707-82-1

Vol. 64-Cool Stars, Stellar Systems, and the Sun, Eighth Cambridge Workshop
ed. Jean-Pierre Caillault   ISBN 0-937707-83-X

Vol. 65-Clouds, Cores, and Low Mass Stars
ed. Dan P. Clemens and Richard Barvainis
ISBN 0-937707-84-8

Vol. 66- Physics of the Gaseous and Stellar Disks of the Galaxy
ed. Ivan R. King
ISBN 0-937707-85-6

Vol. 67-Unveiling Large-Scale Structures Behind the Milky Way
ed. C. Balkowski and R. C. Kraan-Korteweg
ISBN 0-937707-86-4

Vol. 68-Solar Active Region Evolution: Comparing Models with Observations
ed. K. S. Balasubramaniam and George W. Simon
ISBN 0-937707-87-2

Vol. 69-Reverberation Mapping of the Broad-Line Region in Active Galactic Nuclei
ed. P. M. Gondhalekar, K. Horne, and B. M. Peterson
ISBN 0-937707-88-0

Vol. 70-Groups of Galaxies
ed. Otto G. Richter and Kirk Borne
ISBN 0-937707-89-9

Inquiries concerning these volumes should be directed to the:
Astronomical Society of the Pacific
CONFERENCE SERIES
390 Ashton Avenue
San Francisco, CA 94112-1722
415-337-1100
asp @ stars.sfsu.edu

ASTRONOMICAL SOCIETY OF THE PACIFIC
CONFERENCE SERIES

Volume 70

GROUPS OF GALAXIES

Proceedings of a Conference held at the
Space Telescope Science Institute
Baltimore, Maryland USA
June 1992

Edited by
Otto-G. Richter and Kirk Borne

Table of Contents

Foreword	xi

Otto-G. Richter: Groups of Galaxies. Setting the Stage	1
Ben Moore: Selecting Groups of Galaxies in Redshift Catalogues: Results from the CFA Survey	5
Jaime Perea, Ascensión del Olmo, and Mariano Moles: Analysis of Nearby Groups of Galaxies	19
James M. Gelb: Groups of Galaxies in CDM Universes	29
Bradley C. Whitmore: What Determines the Morphological Fractions in Groups and Clusters ?	41
Jane C. Charlton, Bradley C. Whitmore, and Diane M. Gilmore: Pairs in Groups and Clusters	49
Douglas L. Tucker: Groups in the Las Campanas Deep Redshift Survey: A First Look	59
Dennis Zaritsky: Comments on the Distribution of Mass Within Dark Matter Halos	65
Stéphanie Côté and Ken Freeman: Faint Dwarf Galaxies in the Sculptor and Centaurus A Groups	75
B.A. Williams and J.H. van Gorkom: HI Mapping of Compact Groups	77
Gary A. Mamon: Compact Groups: Observations and Theories	83
D. Bettoni, L.M. Buson, L. Maira, and F. Bertola: Dynamics of Early-Type Galaxies in Hickson Compact Groups	95
D. Bettoni and G. Fasano: Morphology of Early–Type Galaxies in Compact Groups	101
Mariano Moles, Ascensión del Olmo, and Jaime Perea: Star Formation and Merging in Hickson Compact Groups	107

Contents

Ascensión del Olmo, Mariano Moles, and Jaime Perea: The Shakbazyan compact Groups and Their Populations — 117

Jack W. Sulentic and Carlos R. Rabaça: Compact Groups of Galaxies: The OLF and its Implications — 127

Stephen E. Zepf: Galaxies in Compact Groups — 135

E. Athanassoula and J. Makino: Simulations of Compact Groups of Galaxies: Some Preliminary Results — 143

Kirk D. Borne and Harold F. Levison: Group Simulations: Looking for Compact Groups — 151

W.K. Huchtmeier and E.D. Skillman: A Search for new dwarf members of the M81 group in the 21cm line of HI — 155

Esther L. Zirbel: Properties of Radio Groups — 161

Gary A. Mamon: Workshop Summary — 173

Foreword

On June 16-18, 1992, the Space Telescope Science Institute hosted one in a series of workshops organized by several "journal clubs". The members of the galaxies journal club had overwhelmingly supported the idea to hold a first workshop ever on groups of galaxies. The relatively small number of participants allowed for in-depth discussions, and the environment at STScI led to many additional private discussions outside the sessions.

Groups of galaxies are interesting for many reasons. Some of the currently active issues are the definition of groups (i.e. the interloper problem), their dark matter content, their dynamical evolution, the merger rate, the over-merging problem, differences between nearby and distant groups, differences between group members and 'field' galaxies, the place of groups in the hierarchy of galaxy aggregates, etc. Naturally, only a subset of all areas of group research can be represented in a volume of this nature. It is hoped that it provides the readers with a sufficiently broad entry into the field to find their way through the large-bodied literature accumulating on groups of galaxies.

Even though the organizers had aimed at a small workshop-type meeting, a lot of preparatory work was needed. Practically all the "behind the scenes" work was done by Sheryl Falgout. Some financial support from STScI was available for travel support of participants and is gratefully acknowledged. Special thanks are extended to Shireen Gonzaga for invaluable assistance in the final preparation of these proceedings. The editors would also like to thank Brad Whitmore for his keen interest and strong support.

Further thanks are due to D.H. McNamara who graciously accepted these proceedings for publication in the ASP series, even though several unforeseen circumstances substantially delayed the submission of the manuscripts to him.

Otto-G. Richter
Kirk D. Borne
Baltimore
1994

GROUPS OF GALAXIES
Setting the Stage

OTTO-G. RICHTER
Space Telescope Science Institute
3700 San Martin Drive, Baltimore, MD 21218, USA

1. INTRODUCTION

More than 200 years ago the French astronomer Charles Messier cataloged the first aggregate of nebulae we now call galaxies. This was, of course, the famous Virgo cluster. The extragalactic nature of the majority of these nebulae was established by Edwin Hubble in the 1920s. At that time already Hubble referred to a "Local Group" of Galaxies. While galaxy clusters soon became a major focus of intense observational and theoretical research research, not much attention was paid to galaxy groups. Only one catalog (by Holmberg, 1937) appeared until the mid-60's.

The seminal articles by Erik Holmberg (1969) and in particular by Gerard de Vaucouleurs (in the last volume of *Stars and Stellar Systems*, 1975, but first written in 1965) brought groups back into the limelight. Now several catalogs of groups appeared (e.g., Turner and Gott 1976, Materne, 1978, 1979, Tully, 1980) first culminating in the work by Huchra and Geller (1982) and Geller and Huchra (1983a,b). Newer catalogs were produced e.g., by Tully and Fisher (1987; see also Tully, 1987), Maia *et al.* (1989) and Ramella *et al.* (1990). A special subcategory of *compact* groups was first surveyed by Rose (1977) and firmly established by Hickson (1982), who just recently issued an atlas of this sub-class of groups (Hickson, 1993).

Groups of galaxies have figured prominently in several areas of extragalactic research ever since. However, this workshop is believed to be the first one truly dedicated to groups and their properties.

2. WHAT IS A GROUP ?

Instead of starting from some statistical prescription on how to find groups in galaxy (redshift) catalogs I prefer to look "from the inside out". Note that there are really two parts to this problem: (a) how do we *define* a group, and, then, (b) how do we *identify* such groups in galaxy catalogs or on sky maps. In other words, we must distinguish between physical and operational definitions of galaxy groups.

How do we traditionally define the Local Group (LG) ? A necessary criterion always applied is that of distance. Galaxies within about a 1 Mpc radius around our own Milky Way are usually considered members of the LG. Other, secondary, criteria like small (or even negative) radial velocity, or resolution into stars are sometimes used but aren't generally needed. A little over 30 galaxies are now thought to belong to the Local Group. The reader should consult van den Bergh (1989; some updated data in van den Bergh, 1992) for a comprehensive discussion. Although the largest fraction of the mass of the Local Group is contained in M 31 and our Galaxy the galaxy content is dominated by relatively faint dwarf irregular and dwarf elliptical galaxies, many of which are satellites of one of the two major galaxies. These faint galaxies undoubtedly exist also in distant groups, but are very difficult to find there because of their low total and surface brightnesses.

Groups "live" in the grey zone between double and multiple galaxies on one hand and clusters of galaxies on the other. Both demarkations are fuzzy at best. Clusters have been intensely studied and several major catalogs are available (Abell 1956, Zwicky et al. 1963-1968, Abell et al. 1988). By the time a galaxy aggregate is found to contain about 50 or more galaxies we generally agree to call it a cluster. Yet, many smaller aggregates are called clusters, notably the Morgan poor clusters (Morgan et al. 1975, Albert et al. 1977). The existence of X-ray radiation and, hence, of an intergalactic medium usually makes us call the aggregate a cluster. An interesting exception from this rule is the group around NGC 1961, whose HI content and X-ray flux was studied by Shostak et al. (1982). Occasionaly, very large and loose groups are being called galaxy clouds.

3. WHY STUDY GROUPS ?

Aside from natural curiosity there are some very good reasons to study groups of galaxies. Since they occupy an intermediate position in the spectrum of galaxy aggregates they will hold important clues to the formation of structure in the universe. Also, they are more numerous than clusters and might, therefore, contain a comparable fraction of the mass in the universe. Determining the fraction of galaxies in groups will lead to a much better understanding of "field" galaxies.

Another important aspect of group studies is the simulation of galaxy encounters. Far fewer particles that in a cluster simulation need to be included, and, hence, better resolution can be used. If we cannot simulate groups in a realistic manner how can we ever hope to simulate clusters properly ?

Last, but not least, groups are useful for distance determinations in the same way that clusters are since all their member galaxies will be at practically identical distances and statistical errors will be reduced (cf. Richter and Huchtmeir, 1984).

4. A QUICK PRIMER

If we move beyond the Local Group out into the realm of what has come to be known as the Local Supercluster of galaxies we first encounter other groups similar to our Local Group, notably the M 81 group. The Virgo cluster as the purported center of our Local Supercluster is the only true cluster within it, although some researches prefer to call the Ursa Major aggregate - often referred to as cloud - also a cluster. It appears that most of the groups we know in the nearby universe populate the *periphery* of a supercluster. According to Fig. 3 in deVaucouleurs (1975) our Local Superclusters appears to be flattened. Conversely, galaxy groups can be an excellent tracer of the structures of hierarchically higher ranking aggregates.

Even for the nearby groups there is no general agreement on their detailed galaxy content and their boundaries as illustrated by Table 4 in Huchra and Geller (1982). Therefore, the definition of a group is by no means an agreed upon procedure. Yet, these nearby groups provide important stepping stones for many of our modern distance estimators. It is beyond the scope of this workshop to address the distance scale discrepancy, but only better distances to such nearby entities will ultimately resolve it. In this field the Hubble Space Telescope will make major contributions once repaired at the end of 1993.

From the first comprehensive analysis of dynamical parameters of galaxy groups by Huchra and Geller (1982) we know that their velocity dispersions are generally less than 300 kms^{-1}, their sizes less than 1.5 Mpc, and their crossing times less than half the age of the universe. Because of small number statistics these numbers tend to be far less accurate than corresponding quantities for galaxy clusters, not the least due to the larger interloper problem. These problems prompted several researches to rediscuss the identification of groups in the original CfA redshift catalog (Huchra et al. 1983). The most recent such discussion by Nolthenius (1992) revised downward the fraction of galaxies within groups, their velocity dispersions and mass-to-light ratios, and interloper contamination while slightly increasing their crossing times.

Before letting everybody else present their results on galaxy groups I like note that the references given below encompass more than has been cited herein. They can be used as a further introduction to the subject.

Acknowledgement: It is a pleasure (a) to thank the Space Telescope Science Institute for hosting this workshop, and (b) its participants for enlightening discussions and active participation.

5. REFERENCES

Abell, G. 1958, *ApJS*, **3**, 211.
Abell, G.O., Corwin, H.G. Jr., and Olowin, . 1988, *ApJS*, , .
Albert, C.E., White, R.A., and Morgan, W.W. 1977, *ApJ*, **211**, 309.
de Vaucouleurs,G. 1975, in *Stars and Stellar Systems*, Vol. **IX**, *Galaxies and the Universe*, eds. A. Sandage, M. Sandage, and J. Kristian, 557.
Geller, M., and Huchra, J. 1983a, *ApJ*, **2xx**, .
Geller, M., and Huchra, J. 1983b, *ApJS*, **52**, 61.
Hickson, P. 1982, *ApJ*, **255**, 382.
Hickson, P. 1993 *Ap&SS*, , .
Holmberg, E., 1937, *Ann.Obs.Lund*, No. **6**, 1.
Holmberg, E., 1969, *Arkiv för Astronomi*, **5**, 305.
Huchra, J., and Geller,M. 1982, *ApJ*, **257**, 423.
Huchra, J., Davis, M., Latham, D., and Tonry, J. 1983, *ApJS*, **52**, 89.
Kraan-Korteweg, R.C., and Tammann, G.A. 1979, *Astron.Nachr.*, **300**, 181.
Maia, M.A.G., daCosta, L.N., and Latham, D.W. 1989, *ApJS*, **69**, 809.
Materne, J. 1978, *A&A*, **63**, 401.
Materne, J. 1979, *A&A*, **74**, 235.
Materne, J., and Tammann, G.A. 1974, *A&A*, **33**, 451.
Morgan, W.W., Kayser, S., and White, R.A. 1975, *ApJ*, **199**, 545.
Nolthenius, R.A. 1992, *ApJ*, , .
Ramella, M., Geller, M.J., and Huchra, J.P. 1989, *ApJ*, **344**, 57.

Richter, O.-G., and Huchtmeier, W.K. 1984, *A&A*, **132**, 253.
Rood, H.J., Rothman, V.C.A., and Turnrose, B.E. 1970, *ApJ*, **162**, 411.
Rose, J.A. 1977, *ApJ*, **211**, 311.
Sandage, A., Binggeli, B., and Tammann, G.A. 1985, *AJ*, **90**, 1759.
Shostak, G.S., 1982, *A&A*, **115**, 293.
Tully, R.B. 1980, *ApJ*, **237**, 390.
Tully, R.B. 1987, *ApJ*, **321**, 280.
Tully, R.B., and Fisher, J.R. 1987, *Nearby Galaxy Catalog*, Cambridge University Press.
Turner, E.L., and Gott III, J.R. 1976, *ApJS*, **32**, 409.
van den Bergh, S. 1989, *A&A Rev.*, **1**, 111.
van den Bergh, S. 1992, *MNRAS*, **255**, 29P.
Zwicky, F., Herzog, E., Wild, P., Karpowicz, M., and Cowal, C.T. 1961-1968, *Catalogue of Galaxies and Clusters of Galaxies*, Vols. 1-6.

Selecting Groups of Galaxies in Redshift Catalogues: Results from the CFA Survey

BEN MOORE

Astronomy Department, University of California, Berkeley, CA 94720, USA

1. INTRODUCTION

Although the Universe is rich in structure over a very large range of scales, most cosmological studies focus either on bright galaxies or on rich clusters. There are practical reasons for this: magnitude limited surveys tend to pick out galaxies in a narrow range of luminosities while the identification of galaxy aggregates in projection requires the high contrast typical of the richest clusters. Only in recent years have extensive redshift surveys over wide areas of sky made it possible to identify intrinsically poor groups of galaxies which produce quite weak enhancements in the projected galaxy distribution. These groups probe scales intermediate between individual bright galaxies and rich clusters, and so may give clues as to the relation of galaxy formation to that of large scale structure. Huchra and Geller (1982) and Geller and Huchra (1983) constructed the first catalogues of groups identified as number density enhancements over the local background. They concentrated on estimates of mass-to-light ratios. The methods introduced by Geller and Huchra were refined and extended by Nolthenius and White (1987; hereafter NW) who used N-body simulations to help determine the optimal group selection procedure in the CfA survey. NW reconsidered the issue of mass-to-light ratios and discussed various other properties of groups both in the CfA survey and in N-body simulations of CDM universes.

We perform a reanalysis of groups in the Center for Astrophysics redshift survey focussing on properties that can be readily compared with predictions of gravitational clustering theories. The grouping algorithm we use is optimised by comparison with artificial catalogues constructed from N-body simulations that have similar low order correlations to those of the original survey. One of our primary aims is to determine the abundance of bound structures as a function of their total luminosity, a "universal" luminosity function of galactic systems which is independent of the detailed arrangement of luminous material within each object. We develop a method for estimating the total luminosity of groups identified in magnitude limited galaxy surveys, and use it to derive a luminosity function of galactic systems. In addition to the abundance of groups by luminosity, we also consider their abundance by velocity dispersion and their luminosity-velocity dispersion relation. Whereas we observe a smooth transition in the luminosity function from single galaxies to luminous rich clusters, the distribution of velocity dispersions shows a discontinuity in slope at the transition between groups and rich Abell clusters. However, this distribution joins smoothly with the distribution of cluster masses inferred from x-ray temperature measurements. We suggest that this is a result of contamination and projection effects artificially inflating the velocity dispersions of many rich clusters. We adopt the Hubble constant, $H_o = 100$ h km s^{-1} Mpc^{-1} with h=0.5.

2. GROUP SELECTION

Groups are defined as aggregates of three or more galaxies picked out by a friends-of-friends or percolation grouping algorithm. This is roughly equivalent to picking objects whose density contrast relative to the local mean is above some chosen minimum. The linking length in the friends-of-friends algorithm must allow for the facts that the galaxy sample is magnitude limited and that redshift separations rather than true distance separations are known. We can see how the procedure works as follows. Suppose first that true distance separations are known. Then to obtain groups approximately bounded by isodensity contours, the linking parameter, r_L, should scale inversely with the one-third power of the local density:

$$r_L \propto \left(\int_{L_{lim}}^{\infty} \Phi(L) dL \right)^{-1/3} \qquad (1)$$

where $\Phi(L)$ is the galaxy luminosity function of the sample and $L_{lim}(z)$ is the luminosity corresponding to the apparent magnitude limit at each redshift.

To investigate the limitations and performance of our grouping algorithm we employ catalogues constructed from N-body simulations evolved from cold dark matter initial conditions. Such catalogues have similar low order correlations to those of the CfA survey. They therefore serve the dual purpose of providing the desired benchmark, while at the same time enabling a direct comparison of theory with observation. Note that in the former role their utility is essentially independent of the validity of the CDM model. Before describing our group finding procedure we briefly discuss the way in which our artificial "CfA" catalogues are generated. We used the five CDM N-body simulations described by Frenk et al. (1990) to extract "galaxy" catalogues analogous to the CfA survey. Each simulation represents a comoving cube of a flat universe of present size 360 Mpc (for h=0.5) containing 262144 particles. Results are analysed assuming three different values for the amplitude of the primordial fluctuation spectrum, corresponding to values of the biasing parameter b=1.6, 2.0, and 2.5. "Galaxies" are identified in the simulations using the "high peak" model as described and implemented by White et al. (1987b) and Frenk et al. (1990).

A slice of declination $0° < \delta < 35°$ through an artificial "CfA" catalogue is illustrated in the "pie" diagram of Figure 1a, in which *real distance* is plotted as the radial coordinate and right ascension as the angular coordinate. Groups identified at an overdensity of about 35, in real space using equation (1), are indicated in Figures 1a and 1b by joining the galaxies linked together by our group finding algorithm. Figure 1b shows the same slice as Figure 1a, but with redshift distance, V, plotted as the radial coordinate. Comparison of these two plots shows the way in which groups are distorted in velocity space. The strategy we adopt below is to tune our group finder so that the groups found in velocity space resemble those found in real space as much as possible. Gravitationally bound associations show up as clumps which are elongated towards the observer and give rise to prominent "fingers of God", but groups with smaller velocity dispersions are much harder to identify visually. The purpose of a group finding algorithm is to pick out these elongated structures taking into account the variation in galaxy number density with distance. It is clear from Figures 1a and 1b that constructing such an algorithm is not straightforward. However, artificial galaxy catalogues in which the distance to each galaxy is known, in addition to its redshift, provide a valuable benchmark against which the performance of a given algorithm can be assessed.

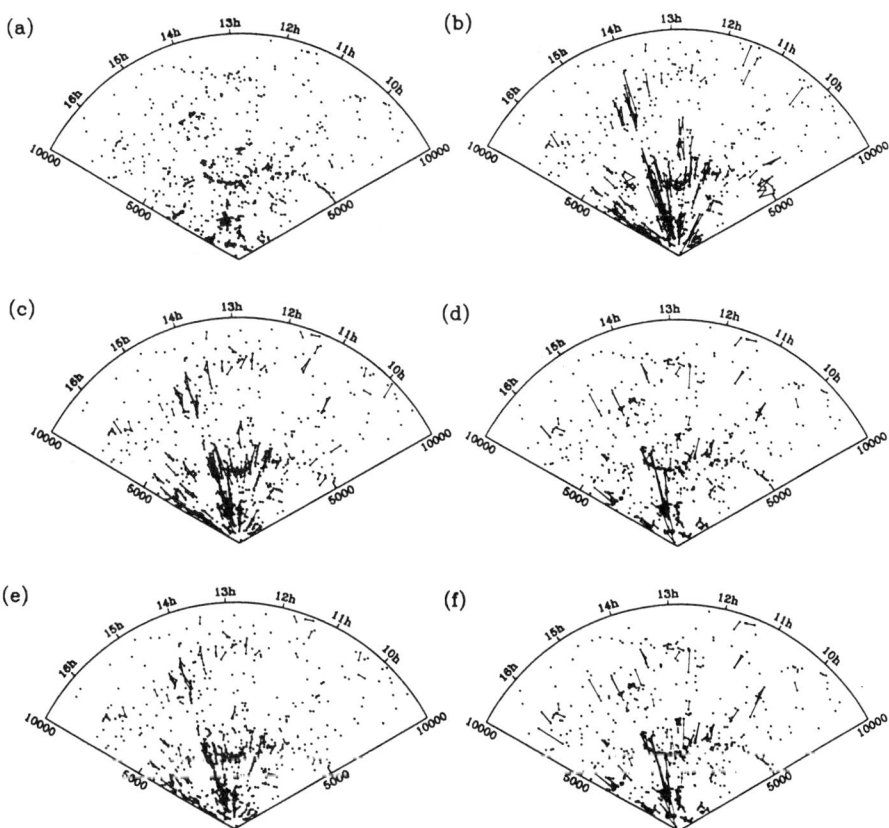

Figure 1. The galaxy distributions of one of our artificial "CfA" catalogues with groups indicated by linking member galaxies together. *(a)* Groups identified in real space and plotted in real space. *(b)* Groups identified in real space and plotted in redshift space. *(c)* Groups identified in redshift space with our standard parameters and plotted in redshift space. *(d)* Groups identified as in (c) plotted in real space. *(e)* Groups identified in redshift space with $V_0 = 350 \ km \ s^{-1}$ and $D_0 = 1.51$ Mpc ($h = 0.5$) plotted in redshift space. *(f)* Groups identified in redshift space with $V_0 = 750 \ km \ s^{-1}$ plotted in real space.

When the depth information comes from the observed velocities, this scaling must be modified to take into account the smearing of positions along the line-of sight. Let us denote the linking cut-offs in angular separation and in line-of-sight velocity by D_L and V_L respectively. NW suggest that D_L should be taken to scale as the inverse square root of the group's surface density. Since, they argue, the luminosity of a typical group scales as the cube of its characteristic size and as the square of its redshift distance, this condition leads to

$$D_L = D_0 \left(\frac{\int_{L_{lim}}^{\infty} \Phi(L) dL}{\int_{L_F}^{\infty} \Phi(L) dL} \right)^{-1/2} \left(\frac{V}{V_F} \right)^{-1/3}, \qquad (2)$$

where V_F is a fiducial recession velocity which we take to be 1250 $km\ s^{-1}$ and L_F is the limiting luminosity at this redshift. NW also argue that V_L should scale as the velocity dispersion of a typical group picked out at each distance since the luminosity and presumably, therefore, the velocity dispersion of the groups picked out increases with distance. The expected variation of group velocity dispersion with distance can be estimated by examining groups in artificial catalogues constructed in *real* space. In agreement with NW, we find that this procedure leads to the following scaling for V_L: $V_L = V_0 + 0.03(V - 5000)$km s^{-1} where V is the mean velocity of the two galaxies to be linked.

There are two free parameters, D_0 and V_0. Many of the properties of interest here are fairly insensitive to D_0. For the most part, we adopt $D_0 = 1.51$ Mpc ($h = 0.5$) or 0.3 times the mean intergalaxy separation at V_F. The choice of V_0 is more difficult. Too small a value excludes galaxies with large relative velocities, artificially depressing the measured velocity dispersion. Too large a value, on the other hand, links non-members into the group, artificially increasing its velocity dispersion. We proceed by comparing artificial group catalogues in real and redshift space and adjust V_L until we obtain the best agreement between the median values of the velocity dispersion. We find that the appropriate value of V_0 needed to match the real space data is somewhere between 550 and 750 km s^{-1}. In this range, the number of galaxies grouped in common between the real and redshift space catalogues reaches a maximum of $\sim 80\%$.

A further constraint on V_0 may be imposed by requiring that the group mass-to-light ratio, M/L, be independent of distance from the observer. We discuss below our methods for estimating group masses and luminosities, but we note here that in both the artificial and the real catalogues, a constant M/L is obtained only for $V_0 \simeq 550$ km s^{-1}. Larger values lead to a systematic increase of this ratio with distance whereas lower values lead to a systematic decrease.

We now illustrate graphically the performance and limitations of our procedure. In Figure 1c we show the groups actually found by our algorithm in redshift space, using our adopted parameter values. Comparison with Figure 1b illustrates how well our procedure works. Most of the "true" groups in Figure 1b are also identified in redshift space, although their detailed memberships are not identical. In Figure 1d we show the groups found in redshift space, but now plotted in real space. This should be compared with Figure 1a. The majority of the groups are found in both plots, but residual contamination by non-members is apparent in Figure 1d. Figure 1e shows groups found in redshift space with our adopted value of D_0, but with $V_0 = 350\ km\ s^{-1}$, plotted in redshift space. Finally, in Figures 1f we plot groups in real space found in redshift space with

$V_0 = 750~km~s^{-1}$. From these last two plots we notice that if V_0 is too low, large groups are artificially subdivided, and if V_0 is too large, the contamination by outlyers becomes prohibitively large. Our adopted value $V_0 = 550~km~s^{-1}$ is a compromise between these two undesirable extremes. In our opinion, this is the best that can be done given the limitation of working with redshifts rather than distances.

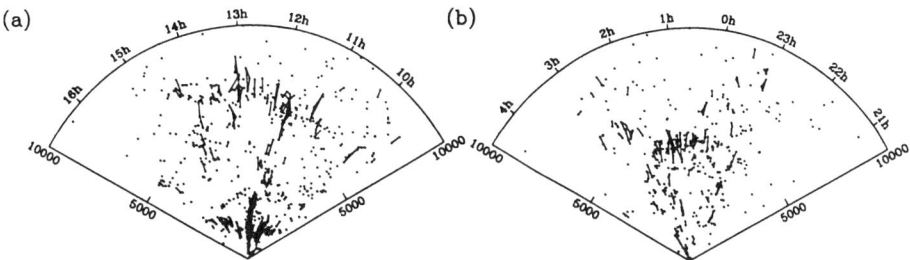

Figure 2. *(a)* Groups in the Northern CfA slice shown in Figure *1(a)*. *(b)* All the groups identified in the Southern CfA survey. Groups were identified using our standard parameters and are indicated by linking member galaxies together.

Figure 2a shows the groups identified in a slice of the Nothern CfA survey, while Figure 2b shows all the groups in the South. The most visible cluster in the North is Virgo which has ~ 200 members and extends out to 3000 km s^{-1}; Coma and Abell 1367 can be also be seen at ~ 7000 km s^{-1}. There are fewer clusters in the South with several of them concentrated in a large supercluster at $\sim 5000~km~s^{-1}$. Over the entire CfA survey we find 166 groups with ≥ 3 members, which include 55% of all galaxies; 13% of galaxies lie in pairs and 32% are singles. The median density contrast of the groups is 50, the median line-of-sight velocity dispersion is 155 km s^{-1}, and the median crossing time is 43% of the Hubble time. (See NW for the exact definitions of these quantities and for a discussion of other related properties of the CfA groups.) *Note that properties derived from the size and velocity dispersion of groups are strongly dependent on the selection cut-offs adopted.*

The angular linking parameter, D_0, is less strongly constrained. It determines the typical density contrast of the groups. In Table 1 we show how the CfA group properties vary as this parameter is changed, so as to obtain group catalogues with density contrasts ranging between 5 and 644. The third column in this table gives the percentage of galaxies grouped in each case, the fourth column gives the number of groups with ≥ 3 members, and the fifth column gives median velocity dispersions. The remaining columns will be discussed below.

3. THE LUMINOSITY FUNCTION OF GALACTIC SYSTEMS

3.1. Groups in the CfA Survey

In this section we discuss our method for estimating the *total* luminosity, L_{tot}, of groups identified in magnitude limited surveys. We begin by assuming a

TABLE 1. Properties of CfA groups

D_L/Mpc Mpc	$(d\rho/\rho)_{med}$	%$_{grouped}$	n_g	σ_{med} km s^{-1}	M/L_z	β	γ	M_z^*	ϕ^* (10 Mpc)$^{-3}$
3.0	5	77	147	178	199	1.04	1.27	−22.1	0.164
2.5	12	71	151	169	181	1.19	1.43	−22.7	0.143
2.0	30	64	169	161	176	1.26	2.62	−22.8	0.175
1.5	50	55	166	155	149	1.34	2.89	−22.9	0.158
1.0	154	49	145	141	123	1.50	3.28	−23.1	0.130
0.5	644	33	110	132	96	1.66	3.90	−23.4	0.080

universal galaxy luminosity function, $\Phi(L)$. The observed galaxies in each group are then a random sample of that portion of $\Phi(L)$ which lies above the local magnitude limit. Note that even single galaxies must be regarded as the sole member above the apparent magnitude threshold of a group. We refer to the distribution of L_{tot} as the luminosity function of groups or "galactic systems." Let L_{obs} be the luminosity of galaxies included in the catalogue and L_{cor} the luminosity of galaxies fainter than the local absolute magnitude limit of the catalogue, $M_{lim} = -5\log v + 5\log H_o - 10.45$. Then, for each group, $L_{tot} = L_{obs} + L_{cor}$. To estimate the contribution of galaxies below threshold we assume that group members are independently drawn from a group luminosity function with a universal Schechter form, $\Phi(L)dL \propto L^\alpha exp(-L/L^*)dL$. We normalise this function to give the observed number of galaxies in each group. The expected luminosity of faint group members is then

$$L_{cor} = N_{obs} \frac{\int_0^{L_{lim}} L\Phi(L)dL}{\int_{L_{lim}}^\infty \Phi(L)dL} \qquad (3)$$

where, $L_{lim} = 10^{0.4 \times (M_\odot - M_{lim})}$. Again, we adopt parameters for the luminosity function derived by Davis and Huchra (1982) from the CfA catalogue, $\alpha = -0.9$ and $L_* = 3.3 \times 10^{10} L_\odot$ (h=0.5). We have checked that our results are insensitive to this choice.

Out to the typical group distance of 5000 km s^{-1}, the mean add on luminosity, L_{cor}, is relatively small; beyond that it increases rapidly. At 5000 km s^{-1}, the add on luminosity per galaxy is $\sim 0.5 L_*$; at 9000 km s^{-1} it has increased to $7 L_*$. At 7000 km s^{-1}, the correction is 100% and for most of our analysis, we shall consider only galaxies up to a distance of 7000 km s^{-1}, where the correction for a single galaxy, $\sim 1 L_*$, is still reasonably small. We shall, however, examine the effect of varying this cutoff in some of the tests described below.

An immediate test of our procedure so far is to compare our values of L_{tot} for Virgo, Coma and Abell 1367 with direct estimates of their total luminosity from deep photometry. At the distance of Virgo, our correction is quite small and it is not surprising that we can recover the total luminosity of Virgo as measured by Sandage et al. (1985). A more stringent test of our luminosity correction is provided by more distant clusters. The Coma cluster lies at over 7000 km s^{-1}, where our correction is quite large. Our algorithm finds only 23 galaxies in Coma, from which we estimate $L_{tot} = 7.9 \times 10^{12} L_\odot$, four times the directly observed value. From the sample of 6724 galaxies in Coma of Godwin et al. (1983) complete to $m_b = 21.0$, we find a total luminosity of $L_z = 7.3 \times 10^{12} L_\odot$ in a $2° \times 2°$ square around the core of Coma. Similarly, for Abell 1367, at 6530 km

s^{-1}, we estimate a total luminosity, $L_z = 2.4 \times 10^{12} L_\odot$ (2.5 times the observed value) from 21 galaxies, whereas the total cluster luminosity (down to $m_b = 19.9$) given by the photometry of Godwin and Peach (1982) is $L_z = 2.3 \times 10^{12} L_\odot$.

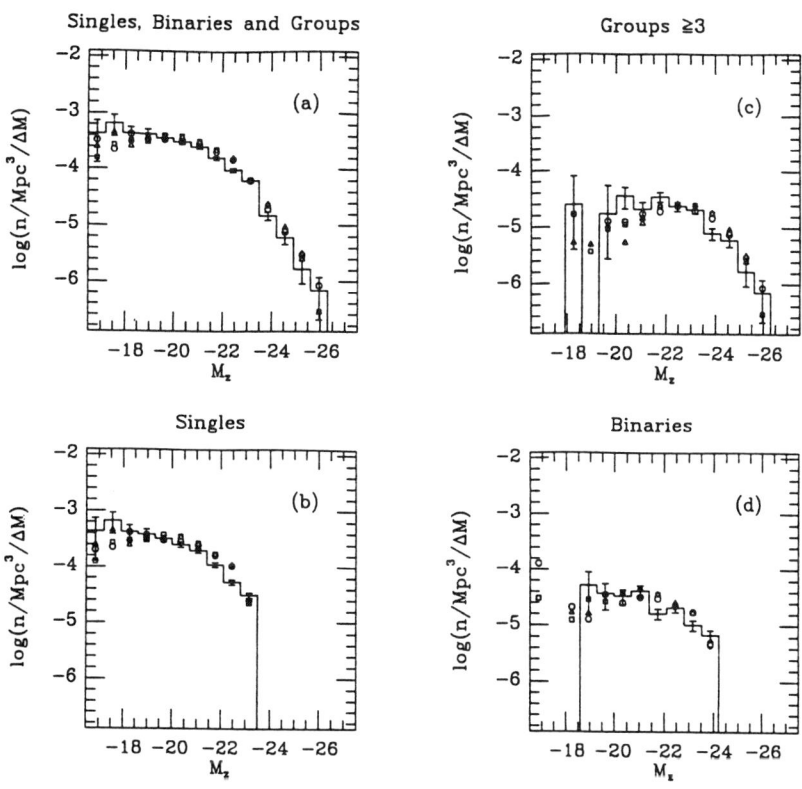

Figure 3. *(a)* Group luminosity functions. *(b), (c), (d)* Contributions to the group luminosity function from groups in which only single galaxies, pairs or 3 or more members are brighter than the magnitude limit. Histograms show the CfA data with Poisson error bars; circles, squares and triangles show the CDM models with biasing parameters $b = 2.5, 2.0$ and 1.6 respectively.

To estimate the group luminosity function, Φ_g, we obtain the space density of groups in the ith absolute magnitude bin by weighting each group by $1/\Gamma_{max}$, where Γ_{max} is the maximum volume within our velocity cutoff out to which the luminosity of the brightest group member is above L_{lim}. Our estimate of the CfA group luminosity function is shown as a histogram in Figure 3a. In Figures 3b-3d, we decompose this luminosity function into separate contributions from groups in which only single galaxies, pairs, or three or more members are brighter than the magnitude limit. The faint end of the group luminosity function is made up almost entirely of single galaxies; binaries link these to the steep bright tail of richer groups and clusters. Group luminosity functions for the CDM groups are shown by symbols in Figure 3. Averages from all the simulations in each ensemble are shown. The CDM group catalogues produce luminosity functions which are in reasonable agreement with that of the CfA groups.

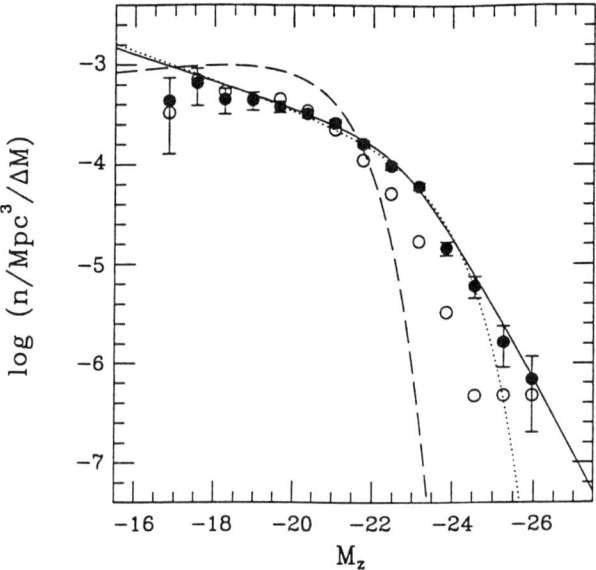

Figure 4. The CfA group luminosity function (filled circles). The open circles show the luminosity function without the correction of equation (3). The solid line shows the best fit double power law to the data, (equation 4), while the dotted lines show the best fit Schechter function. Our adopted field galaxy luminosity function (dashed line) is shown for comparison. The error bars for the filled circles are $\pm 1\sigma$ Poisson errors.

The group luminosity function is reproduced in Figure 4 (filled circles), where we also show the result obtained without applying our luminosity correction (open circles). For comparison, we show the field galaxy luminosity function for the CfA survey. A Schechter function with $\alpha = -1.37$, $M_* = -23.53$ and $\Phi_* = 1.25 \times 10^{-4}$) provides a good fit to the group luminosity function over the range $-24 < M_z < -17$, but is too steep at the bright end (dotted line). We find that a good fit over the entire range plotted is provided by a "Marshall" (1987) double power law,

$$\phi(L)dL = \phi^{*'}[(\frac{L}{L_*})^\beta + (\frac{L}{L_*})^\gamma]^{-1}\frac{dL}{L_*}, \qquad (4)$$

with best-fit parameters $\beta = 1.34 \pm 0.07$, $\gamma = 2.89 \pm 0.12$, $M_z^{*'} = -22.93 \pm 0.2$ and $\phi^{*'} = 1.58 \pm 0.5 \times 10^{-4}$ Mpc^{-3}.

We have argued that our definition of groups as aggregates bounded (approximately) by an equidensity surface at 10–100 times the mean galaxy density is physically reasonable. There is some freedom, however, in choosing grouping parameters to satisfy these criteria and it is important to establish how our results depend on this choice. In Figure 5a we illustrate the effect of varying V_0 while keeping D_0 fixed in our artificial "CfA" group catalogues. The luminosity function is seen to be only weakly sensitive to V_0. Also shown in this figure (as a solid line) is the luminosity function of groups found in the same artificial catalogues but using *real* rather than redshift distances. The luminosity function in real space agrees well with those in redshift space, especially for our adopted value of $V_0 = 550$ km s^{-1}. Note that the extreme values of V_0 adopted in Figure 5a give rise to quite different velocity dispersion distributions, even though the luminosity functions are rather similar. This simply reflects the large weight given to outlying members along the line-of-sight in estimates of velocity dispersion.

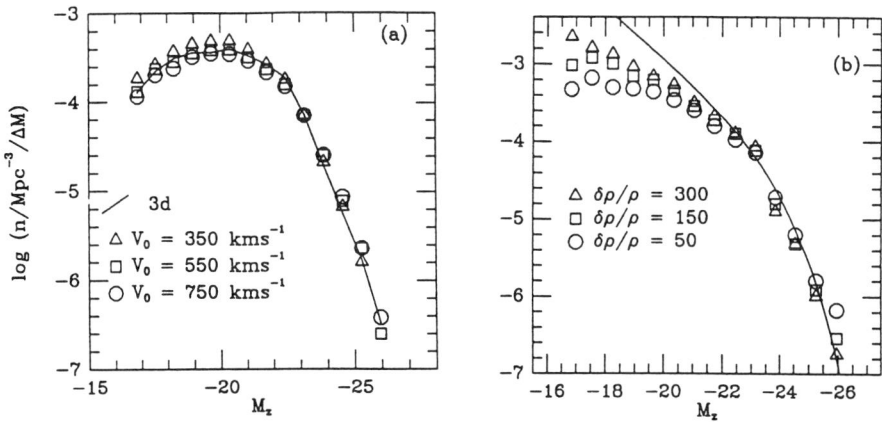

Figure 5. The sensitivity of the group luminosity function to the grouping parameters *(a)* Group luminosity functions in artificial "galaxy" catalogues constructed from N-body simulations. The symbols give results for groups identified in redshift space with the radial linking parameter, V_0, set to 350 km s^{-1} (triangles), 550 km s^{-1} (squares), and 750 km s^{-1} (circles). The solid line shows the luminosity function of groups found in *real* space. *(b)* Luminosity function of groups of different characteristic density contrast. The open symbols show results for CfA groups with median values of $\delta N/N$ equal to 50 (circles), 150 (squares) and 300 (triangles). The solid line is the Press-Schechter prediction for a CDM model with b=2 assuming a constant $(M/L) = 250$.

Varying the angular linking length, D_0, produces group catalogues in which the median density contrast decreases with increasing D_0. We have constructed catalogues with median density contrasts between five and several hundred, by varying D_0 between 3 and 0.5 Mpc while keeping V_0 fixed at 550 km s^{-1}. In

Figure 5b we show luminosity functions for three values of D_0, this time for the CfA groups. In Table 1 we list the best-fit double power-law parameters for our groups catalogues. The characteristic group luminosity, M_*, is insensitive to D_0, but the slopes at the bright and faint ends vary systematically with the median density contrast and thus with the fraction of galaxies grouped. Table 1 gives median values of the virial mass-to-light ratio for these catalogues. Plotted as the solid line in this diagram is a prediction of the Press-Schechter theory (1974), where we have assumed the CDM power spectrum with b=2 and a constant mass to light ratio of 250.

3.2. Further Tests of the Methods

The most unusual aspect of our procedure is to regard single galaxies as "groups" whose other members are too faint to be included in a magnitude limited catalogue. We carried out a series of tests to verify that our results do not depend on the magnitude limit of the catalogue or on the adopted velocity cutoff.

To test the first point we generated artificial galaxy catalogues, magnitude limited at $m_z = 15.5$ from our $b = 2.0$ ensemble of simulations. We compared results for these catalogues with those from our standard ones limited at $m_z = 14.55$ and with subcatalogues extracted from them at a magnitude limit of $m_z = 13.5$. As a separate test we compared group luminosity functions for catalogues limited at $m_z = 14.5$, but with different velocity cutoffs at 10000 km s^{-1}, 7000 km s^{-1} and 4000 km s^{-1}. We found no systematic differences amongst the AGS luminosity functions in these catalogues indicating that our method for estimating L_{tot} is indeed independent of the magnitude limit and choice of velocity cutoff.

4. THE DISTRIBUTION OF GROUP VELOCITY DISPERSION

The cumulative distribution of line-of-sight velocity dispersion, $<V_{l.o.s}^2>^{1/2}$, for CfA groups with three or more members is plotted in Figure 6a. To obtain a true spatial density, each group was weighted by $1/\Gamma_{max}$, the maximum volume in which the group would have been detected. The abundance is normalised to the mean density of Abell clusters of richness class $R > 0$, $n_a = 7.5 \times 10^{-7}$ Mpc^{-3} (Bahcall and Soneira 1983). Also shown in the figure is the corresponding distribution for a sample of over 80 Abell clusters taken from the literature. Finally, the solid line shows the distribution of equivalent velocity dispersion obtained from the distribution of X-ray temperature for rich clusters given by Edge et al. (1990). To convert temperature to velocity dispersion we took $\beta \equiv <V_{l.o.s}^2>/[kT_x(\mu m_p)^{-1}] = 1.25$ as indicated by Evrard's (1990) theoretical models.

There are several noteworthy features in this plot. The abundance of groups falls off very steeply with velocity dispersion. It joins fairly smoothly onto the inferred distribution for X-ray clusters which has a similar slope and continues to larger values of $<V_{l.o.s}^2>^{1/2}$. On the other hand, the abundance of $R > 0$ Abell clusters as a function of directly measured velocity dispersion is rather different. Although at $<V_{l.o.s}^2>^{1/2} \simeq 500$ km s^{-1}, it roughly matches the abundance of CfA groups and X-ray clusters, at larger dispersions it falls off much less rapidly, and there is still a significant population of clusters at $< V_{l.o.s}^2 >^{1/2} \simeq 1500 \; km \; s^{-1}$. The agreement between the CfA groups and the X-ray data, seems to lend further support to the view that many of the cluster velocity dispersion measurements are artificially inflated due to contamination by foreground galaxies.

The predictions of our CDM models are compared with the CfA data in Figure 6b. At small values of $< V_{l.o.s}^2 >^{1/2}$, the predicted distributions depend only weakly on the biasing parameter, b, and tend to be higher than the CfA data. Thus, the median dispersion of 224 km s^{-1} for our standard CDM model is substantially larger than the corresponding 155 km s^{-1} for the CfA groups found with identical linking parameters. This discrepancy was already noted by NW, and may be related to questions of "velocity biasing" (Carlberg, Couchman and Thomas 1990). It may well reflect, at least in part, the fact that the simulations do not have sufficient resolution to take into account the internal degrees of freedom of individual galaxies or their interaction with diffuse dark matter.

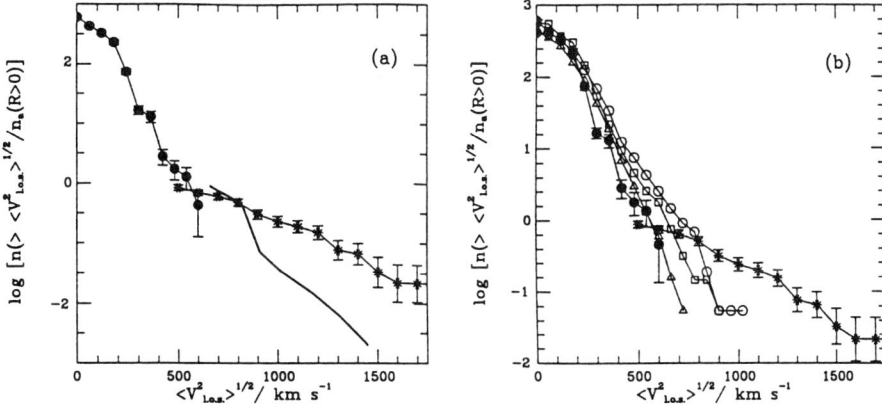

Figure 6. (a) The cumulative distribution of line-of-sight velocity dispersion for galaxy groups and clusters. The circles show the distribution for the CfA groups. The stars show the fractional cumulative distribution for a sample of Abell clusters. The solid line shows the distribution of cluster masses inferred from X-ray temperature measurements. (b) The distributions plotted in (a) together with the corresponding distributions for groups found in CDM catalogues with biasing parameter b=2.5 (triangles), 2.0 (squares), and 1.6 (circles). The abundances are normalised to 7.5×10^{-7} Mpc^{-3}, the observed mean density of Abell clusters of richness class $R > 0$.

In Figure 7 we plot luminosity against velocity dispersion for our CfA (circles) and b=2 CDM (crosses) groups. Since these systems are defined by an overdensity criterion, we expect $< V_{l.o.s}^2 >^{1/2} \propto L^{1/3}$ if the groups are in dynamical equilibrium and have similar mass-to-light ratios. This expectation is approximately born out by both sets of data, even though the scatter is large. The observed trend is consistent with the view that our grouping algorithm identifies real dynamical systems which are not unduly contaminated by unrelated galaxies. For comparison, we also plot the Tully-Fisher relation for spiral galaxies (dashed line; from Aaronson and Mould 1983) and the Faber-Jackson relation for elliptical galaxies (dotted line; from Davies et al. 1983).

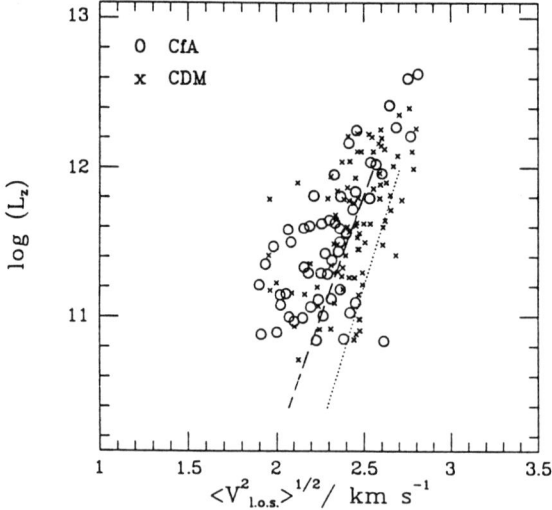

Figure 7. Luminosities of groups with ≥ 5 members plotted against their velocity dispersions. The open circles indicate the CfA groups and the crosses groups from one of our $b=2$ CDM catalogues. For comparison, we also plot the Tully-Fisher relation for spiral galaxies (dashed line) and the Faber-Jackson relation for elliptical galaxies (dotted line).

5. CONCLUSIONS

We have identified groups of galaxies in the Center for Astrophysics (CfA) redshift survey using an algorithm similar to that proposed by Huchra and Geller (1982) and Nolthenius and White (1987), but with somewhat different parameters. We defined groups as galaxy aggregates approximately bounded by isodensity surfaces, with mean overdensity ~ 50. We optimised our grouping algorithm by comparing groups found in redshift space with groups found in real space in artificial "galaxy" catalogues constructed from large N-body simulations. The groups we find have velocity dispersions in the the range $\sim 50 - 600$ $km\,s^{-1}$. They correspond to structures intermediate between bright galaxies and rich galaxy clusters. We examined the continuity of properties over this range of scales and compared the structure and abundance of groups with predictions from the standard cold dark matter cosmogony, obtained using artificial catalogues with the same geometry and magnitude limit as the CfA survey.

We defined the *total* luminosity of a group identified in a flux limited survey as the sum of its observed luminosity plus a suitably normalised integral of the (assumed universal) galaxy luminosity function below the flux limit of the survey. This definition directs us to regard a single galaxy as the sole member of a group lying above the magnitude limit of the catalogue. With our definition and estimation procedures we were able to predict correctly the observed total luminosity of Virgo, Coma and Abell 1367 from the relatively few galaxies above the apparent magnitude limit of the CfA survey. We carried out a variety of tests of our methods using both the CfA data and our N-body models. The luminosity function of galactic systems interpolates smoothly between single bright galaxies at the faint end and luminous rich clusters at the bright end. The luminosity function is quite insensitive to the linking parameter in velocity space – including a few interlopers or leaving out a few genuine members has a relatively minor effect on the inferred group luminosity function (although it may have a large effect on the distribution of velocity dispersion).

The abundance of CfA groups falls steeply with increasing velocity dispersion. Our measured distribution joins smoothly onto that inferred for X-ray emitting rich clusters (which extends out to values ~ 1200 km s^{-1}). However, the abundance-velocity dispersion relation measured directly for Abell clusters has a very different slope and implies a substantially higher abundance of high velocity dispersion clusters. We argued that this situation suggests that the tail of clusters with large measured dispersions may be a result of contamination by foreground galaxies. Our velocity dispersions scale approximately as the 1/3 power of group luminosity, as expected for systems in dynamical equilibrium defined according to an overdensity criterion.

Galaxy groups like those we have considered here are weak enhancements in the overall clustering pattern and so are best identified in redshift surveys. The uncertainties in their identification and in their inferred properties are different from those that are relevant to rich clusters identified in two-dimensional survey data. In both cases the measured velocity dispersions depend sensitively on membership criteria. Other properties, like the luminosity function are more robust. Here we have shown how N-body simulations can be used to optimise the grouping procedure and circumvent some of these limitations. More extensive redshift surveys will enable our techniques to be extended to the study of richer systems.

ACKNOWLEDGEMENTS I would like to thank my collaborators on this project, Carlos Frenk and Simon White, for allowing me to present these results prior to publication.

6. REFERENCES

Abell, G. 1958, *Ap.J. Suppl.*, **3**, 211.
Aaronson, M. and Mould, J. 1983, *Ap.J.*, **265**, 1.
Bahcall, N.A. 1979, *Ap.J.*, **232**, 689.
Bahcall, N. and Soneira, R. 1983, *Ap.J.*, **270**, 20.
Bhavsar, S.P. 1978, *Ap.J.*, **222.**, 412.
Bhavsar, S.P., Gott, J.R. and Aarseth, S.J. 1981, *A.J.*, **246**, 656.
Binggeli, B., Tammann, G.A. and Sandage, A. 1987, *Astr.J.*, **94**, 251.
Carlberg, R.G., Couchman, H.M.P. and Thomas, P.A. 1990, *Ap.J.*, **352**, L29.
Colless, M.M. and Hewett, P. 1987, *M.N.R.A.S.*, **224**, 453.
Davies, R.L., Efstathiou, G., Fall, M.S., Illingworth, G. and Schechter, P.L. 1983, *Ap.J.*, **266**, 41.
Davis, M., Efstathiou, G., Frenk, C.S. and White, S.D.M. 1985, *Ap.J.*, **292**, 371 (DEFW).

Davis, M. and Huchra, J. 1982, *Ap.J.*, **254**, 437.
Davis, M., Huchra, J., Latham, D.W. and Tonry, J. 1982, *Ap.J.*, **253**, 423.
Edge, A.C., Stewart, G.C., Fabian, A.C. and Arnaud, K.A. 1990, *M.N.R.A.S.*, **245**, 559.
Efstathiou, G., Bond, J.R. and White, S.D.M. 1992, *MNRAS*, **258**, 1P.
Efstathiou, G., Ellis, R.S. and Peterson, B. 1988b, *M.N.R.A.S.*, **232**, 431.
Evrard, A.E. 1990, *Ap.J.*, **349**, 188.
Evrard, A.E. Summers, F.J. and Davis, M. 1992, *Berkeley preprint*.
Felten, J.E. 1985, *Comm. Ap. Sp. Sci.*, **11**, 53.
Fisher, J.R. and Tully, R.B. 1981, *Ap.J.*, **47**, 139.
Frenk, C.S., White, S.D.M., Efstathiou, G. and Davis, M. 1988, *Ap.J.*, **327**, 507.
Frenk, C.S. 1989, in **The Epoch of Galaxy Formation**, eds. C.S. Frenk, R.S. Ellis, T. Shanks, A. Heavens and J. Peacock, (Dordrecht: Kluwer Acad. Publ.), in press.
Frenk, C.S., White, S.D.M., Efstathiou, G. and Davis, M. 1990, *Ap.J.*, **351**, 10.
Frenk, C. S. 1991, in **Nobel Symposium No 79: The birth and early evolution of our Universe**, *Physica Scripta*, **T36**, 70.
Geller, M.J. and Huchra, J.P. 1983, *Ap.J.Suppl.*, **52**, 61.
Godwin, J.G., Metcalfe, N. and Peach, J.V. 1983, *M.N.R.A.S.*, **202**, 113.
Godwin, J.G. and Peach, J.V. 1982, *M.N.R.A.S.*, **200**, 733.
Gott, J.R. and Turner, E.L. 1977, *Ap.J.*, **216**, 357.
Henry, P.J. and Arnaud, K.A. 1991, *Ap.J.*, **372**, 410.
Huchra, J.P., Davis, M., Latham, D.W. and Tonry, J. 1983, *Ap.J. Suppl.*, **53**, 89.
Huchra, J.P. and Geller, M.J. 1982, *Ap.J.*, **257**, 423.
Kent, S. M. and Gunn, J.E. 1982, *Astr. J.*, **87**, 945.
Maddox, S.J., Efstathiou, G., Sutherland, W.J. and Loveday, J. 1990, *M.N.R.A.S.*, **242**, 43p.
Marshall, H.L. 1987, *Ap.J.*, **316**, 84.
Nolthenius, R. and White, S.D.M. 1987, *M.N.R.A.S.*, **225**, 505 (NW).
Press, W.H. and Schechter, P. 1974, *Ap.J.*, **187**, 425.
Ramella, M., Geller, M.J. and Huchra, J.P. 1989, *Ap.J.*, **344**, 57.
Sandage, A., Binggeli, B. and Tammann, G.A. 1985, *A.J.*, **90**, 1759.
White, S.D.M. and Rees, M. 1978, *M.N.R.A.S.*, **183**, 341.
White, S.D.M. and Negroponte, J. 1982, *M.N.R.A.S.*, **201**, 401.
White, S.D.M., Frenk, C.S., Davis, M. and Efstathiou, G. 1987b, *Ap.J.*, **313**, 505.
White, S.D.M. and Frenk, C.S. 1991, *Ap.J.*, **379**, 52.
Zabludoff, A.I., Huchra, J.P. and Geller, J.G. 1990, *Ap.J.Supp.*, **74**, 1.

Analysis of Nearby Groups of Galaxies

JAIME PEREA, ASCENSIÓN DEL OLMO, AND MARIANO MOLES
Instituto de Astrofísica de Andalucía,
Apdo. 3004, 18080–Granada, Spain

ABSTRACT: The two well known CfA samples of groups of galaxies are revisited. By properly assessing errors in parameters such as the mass, the velocity dispersion or the typical sizes of the groups, a significant population of interlopers is found. Two subsamples of well defined groups are obtained by taking into account errors and interlopers. The distribution of the group properties and the correlations between parameters for these two samples are analysed. The relations between the density and M/L and between the morphological content and the velocity dispersion indicate that properties of compact groups may be shared by normal groups.

1. INTRODUCTION

In the analysis of systems of galaxies, groups constitute interesting test cases to confront our current knowledge about topics such as the distribution of mass in the Universe, morphological evolution, or the influence of the environment on individual properties of the galaxies. It appears that there is a continuity in the observed properties of the aggregates, such as the velocity dispersion, sizes or the local density–morphological type relation (Postman and Geller, 1984), from rich clusters to poor groups; clusters being richer, denser and probably more evolved. In addition, in the high density tail of galaxy aggregates there lie the population of compact groups, whose interesting observational properties we are just beginning to understand in some detail and, as some studies suggest (e.g. Barnes, 1989), whose origin and dynamical evolution could be connected with that of loose normal groups. Furthermore, groups of galaxies play a fundamental role in the classical problem of the discrepancy between the luminous and dynamical mass in clusters. It is therefore of interest not only to characterize the properties of normal groups but also to use them as calibrators of properties such as the mass, the density or the morphological content and compare them with those of the other two classes of galaxy systems or, perhaps, even to use groups as a link to the properties of pairs and isolated galaxies.

From the point of view of a first overall analysis, normal groups lie at an intermediate stage between clusters and compact groups and they present some advantages when comparing with the other two kinds of systems. In fact, in comparison to clusters, the samples are usually better defined, they are on average nearer and the fraction of members that can be observed per system is larger. When comparing with compact groups, the samples are more numerous and tidal effects are expected to be smaller. However, when comparing to clusters, the derivation of typical properties such as the mass, the characteristic sizes, or the velocity dispersion, will be affected by larger errors due to the small number of members. Therefore, a proper analysis of possible errors and biases which are included in the group samples is of crucial importance.

In this work we present a statistical and detailed analysis of the two well known samples of groups established by Huchra and Geller (HG, 1982) and

Geller and Huchra (GH, 1983). We have selected groups with $3 < N_{gal} \leq 23$ for the HG sample and $4 < N_{gal} \leq 23$ for the GH sample. In order to analyse the statistical properties, we have evaluated for each group the following set of parameters:

- σ_V — Velocity dispersion.
- M_V — Virial mass.
- M_P — Projected mass.
- R_P — Mean pairwise separation, first moment of the interseparation distribution.
- R_H — Harmonic radius, moment of order -1.
- R_2 — Second moment or gyration radius.
- L_T — Total luminosity.
- f_S — Spiral (morphological type > 0) fraction.
- ϵ — Ellipticity.
- t_{cr} — Crossing time.
- M/L — Mass to Luminosity relation.
- ρ_L — Observed density.

The two mass estimators are those defined by Heisler *et al.* (1985) and are evaluated as indicated by Perea *et al.* (1990a). The total luminosities were determined by applying the unrestricted N brightest method defined by Schechter and Press (1976) adopting a value of $\alpha = -1.3$ for the exponent of the Schechter's luminosity function. The ellipticity was calculated following the method described by Thonatt (1989) and for the observed luminous density ρ_L we use the expression derived by Hartwick (1978)

$$\rho_L = \frac{3f_g L}{4\pi R_H R_2^2}$$

being f_g the mean M/L for galaxies, where a value $f_g = 10 M_\odot/L_\odot$ is adopted here. Throughout this paper we will adopt a Hubble constant of $H_0 = 100$ km s^{-1} Mpc^{-1}.

2. ERROR ANALYSIS AND SELECTION EFFECTS

2.1. Errors

When handling small data samples a robust analysis of the statistical uncertainty for the observed parameters in each group of the sample is required. A well defined measure of the error on the determination of a given parameter can be obtained by means of the Jacknife method (see e.g. Mosteller and Tukey, 1977). This technique is based on the determination of a set of pseudovalues (as many as the number of galaxies in the group) of the parameter by evaluating it for all the galaxies but one. The dispersion of these pseudovalues gives the estimation of the error.

For the HG and GH samples we selected, we have derived estimations of the errors on the scale parameters (R_H, R_P, σ_v) and mass estimators (M_V, M_P). In Figure 1 we present the distributions of relative errors in M_V, M_P, R_H and σ_V for the GH sample. In all the cases there is a tail of non well defined groups

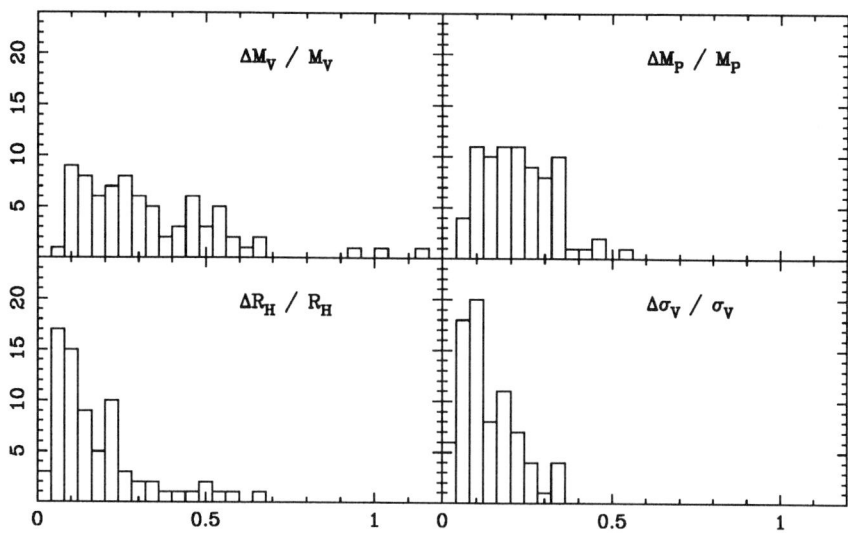

FIGURE 1. Distribution of relative errors for the GH sample.

in the distributions. As it can be clearly seen from Figure 1 the largest errors are measured for the virial mass, mainly induced by R_H which it is known to be ill defined. On the other hand, M_P seems to be less noisy and it provides with relative errors which are significatively smaller.

2.2. Interlopers

One of the most important sources of errors in the determination of dynamical parameters in an aggregate of galaxies is the presence of unbound or projected galaxies and this turns to be crucial for the analysis of small groups. The Jacknife technique is also useful to detect interlopers (Perea et al. 1990b) taking as a basis the influence of each member galaxy in the dynamics of the whole group. We considered a galaxy as being an interloper and thence eliminated from the group if its pseudovalues, as defined in previous section, in M_V and/or M_P depart by more than a factor of 2 of those obtained for the other members. A first consequence of this procedure is the removing of unbound galaxies until the group becomes a pair, all these cases were not included in the forthcoming analysis.

As it is the case for the analysis made for clusters (Perea et al. 1990b), the net effect of the interlopers in the GH and HG samples is a systematic overestimation of the masses of the groups by factors reaching one or two orders of magnitude. This result is clearly illustrated in Figure 2, where both estimators are displayed before and after elimination of the interlopers. As can be observed, M_P is more affected by their presence, however once interlopers are taken into account, smaller errors are obtained for M_P when comparing to those of M_V. This fact shows that while one of the estimators is less sensitive to unbound members, the other is less noisy, thus the comparison between the two mass

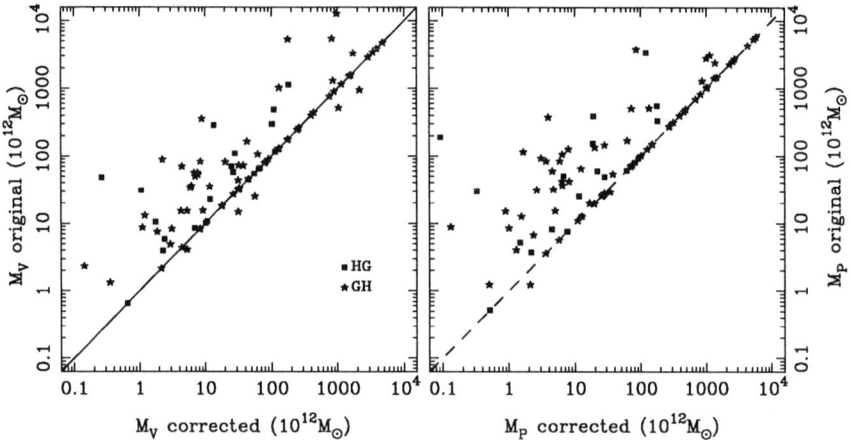

FIGURE 2. The effect of interlopers on the mass determination.

estimators can be very useful when obtaining well defined value for the observed mass of systems of galaxies.

In addition to the elimination of interlopers and in order to obtain a fair sample, we eliminate all the groups with relative errors in M_V and/or M_P greater than the 50%. This leave us with a final population of 20 and 60 groups for the HG and GH samples respectively.

2.3. Biases

There exist important selection effects on the observed parameters for the GH sample. It is noted that farther groups tend to be larger, more massive and have higher velocity dispersions. This fact is shown in Figure 3, where we present the relations with the redshift and the corresponding linear fits, which are parametrized in Table 1 which includes the correlation coefficients and associated probabilities for the log–log relations.

TABLE 1. Correlation coefficients with the redshift and associated probabilities.

Sample	Parameter	σ_V	R_H	M_V	M_P	L	t_{cr}	M/L
GH	r	0.69	0.70	0.77	0.75	0.81	0.03	0.06
	Prob.	1.00	1.00	1.00	1.00	1.00	0.15	0.35
HG	r	0.23	0.05	0.19	0.12	0.70	0.24	0.37
	Prob.	0.67	0.16	0.59	0.39	1.00	0.69	0.89

No trend is found neither for M/L nor for the crossing time, it seems that there is a cancellation of tendencies for both samples between the mass and the luminosity and between the harmonic radius and the velocity dispersion.

With respect to the HG sample, we observe that it does not present any strong bias except on the luminosity (see Table 1). This sample gives typical

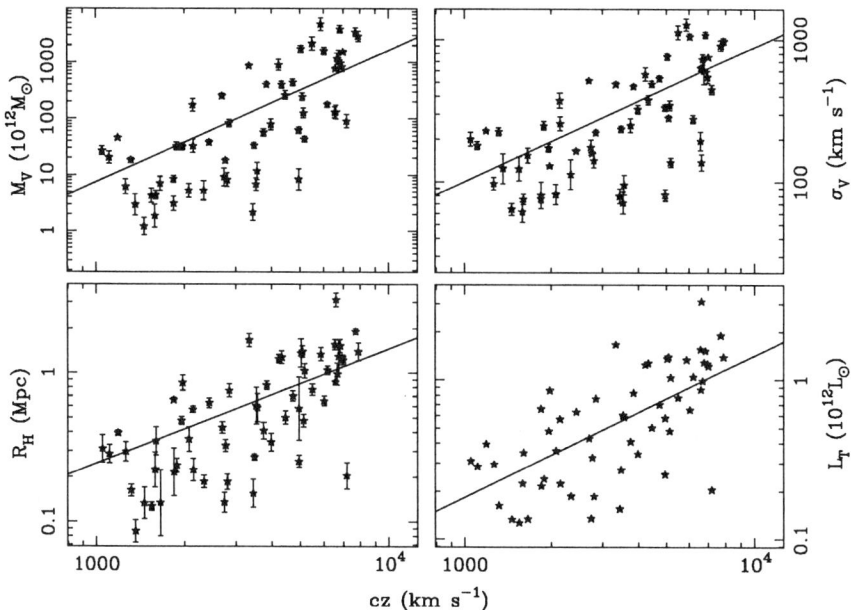

FIGURE 3. Parameters as a function of the redshift for the GH groups.

values for the parameters that are better defined with respect to selection effects, however the smaller number of groups make the statistical errors larger when comparing with the other set of groups. The comparison of the two samples will be needed to balance these two effects.

3. MAIN PROPERTIES

3.1. Distributions

The observed distribution functions of the parameters are not far from being represented by log–normal distributions. The values corresponding to the medians of the log distributions for the two samples are presented in Table 2.

TABLE 2. Central values.

Sample	M_V	M_P	σ_V	ρ_L	R_H	R_P	M/L	t_{cr}
	$10^{12} M_\odot$		$km\,s^{-1}$	$10^{12} M_\odot Mpc^{-3}$	Mpc		$M_\odot/L_\odot$	H_0^{-1}
HG	26	20	154	1.3	0.4	0.8	233	0.64
GH	51	66	231	0.9	0.5	0.9	338	0.32

To test the differences between the two samples, the Kolmogorov–Smirnov and Student's t tests were applied. The results obtained indicate that there exist significant (probability > 95%) statistical differences between the GH and HG samples for both mass estimators, the crossing time and the velocity dispersion. GH groups tend to be more massive and they have shorter crossing times. These discrepancies are mainly caused by the velocity dispersion since the values for R_H or R_P are similar for both cases. Probably the observed differences between the two samples can be explained in terms of the selection criteria plus biases which, as we saw in section 2.3, turn to be very important for the GH groups (see Figure 3).

3.2. Relations between parameters

Once the samples are cleaned from interlopers it is found that for the two samples no clear discrepancy exist between the two mass estimators. In fact and following a Kolmogorov–Smirnov test we observe that M_P and M_V yield the same result at the 98.5% level of confidence, and if any discrepancy exist is in the sense of being the values measured for M_P slightly larger than those of M_V (we found $M_P \approx 1.04 M_V$).

With respect to the relative influence of other parameters on the mass we found that, as it is expected, both mass estimators mainly depend on σ_V. This fact is shown in Table 3 where the correlation coefficients of the mass estimators with several parameters (in log–log form) are presented. A cleaner relation it is found for the mass on the halo size (as measured by R_P) than on the core size (R_H) for HG groups. The lack of correlation of the mass estimators with R_H is mainly due to the larger errors of this last parameter as it was already mentioned in section 2.1.

TABLE 3. Relation of the mass with other parameters.

GH	σ_V	R_H	R_P	L	HG	σ_V	R_H	R_P	L
M_V	0.96	0.78	0.76	0.64		0.96	0.47	0.83	0.50
M_P	0.95	0.74	0.80	0.64		0.95	0.42	0.84	0.45

The mass estimation for systems of galaxies is always based on the assumption of spherical symmetry. In order to analyse the influence that departures from symmetry have on the estimation of the mass, we explored the relation of M_V and M_P with the observed ellipticity of the groups. The result we found is that no trend exists for any of the estimators in the two samples. However it is detected a significant dependence of the relative errors of the mass estimators on the ellipticity (see Figure 4) indicating that the lack of symmetry mainly affect to the quality of the estimator instead to the estimator itself.

An important observed relation is the one which connects the luminous density with the mass to luminosity relation. This result was already found for deVaucouleurs' groups by Hartwick (1978), where a relation of the form $M/L \propto \rho_L^{-1}$ was suggested. It can be seen from the definitions of M/L and ρ_L that the previous expression is equivalent to $R^2 \propto \sigma_V$. These trends clearly appear in the two CfA samples of groups as it is shown in Figure 5, and are of the same kind of those observed by Hartwick. Denser groups tend to have smaller M/L, indicating that there exists a continuity of properties between

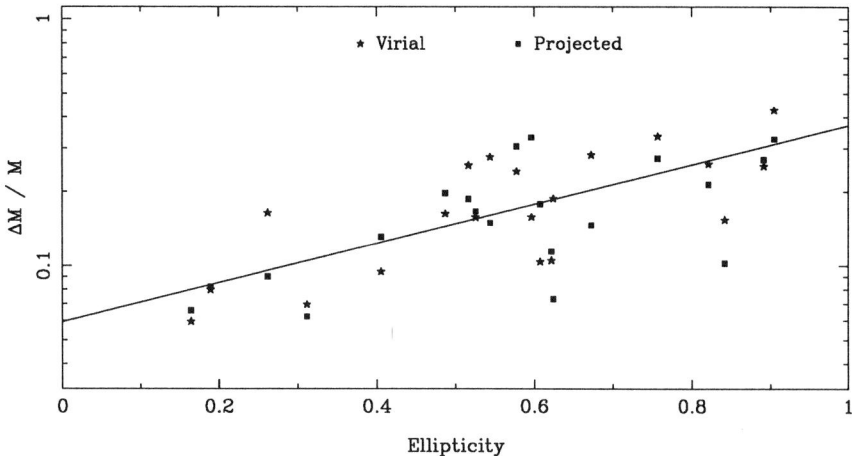

FIGURE 4. Relative errors for the mass estimators as a function of the ellipticity, the fit corresponds to the virial estimator.

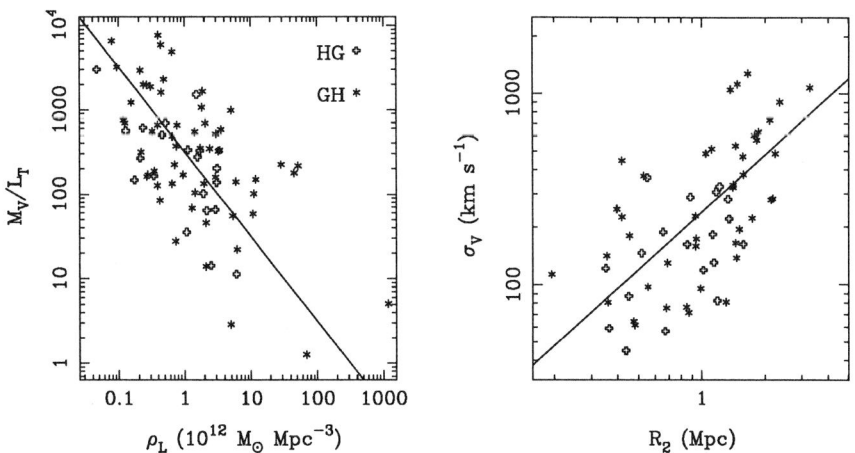

FIGURE 5. M/L vs. the density, and the size–σ_V relation.

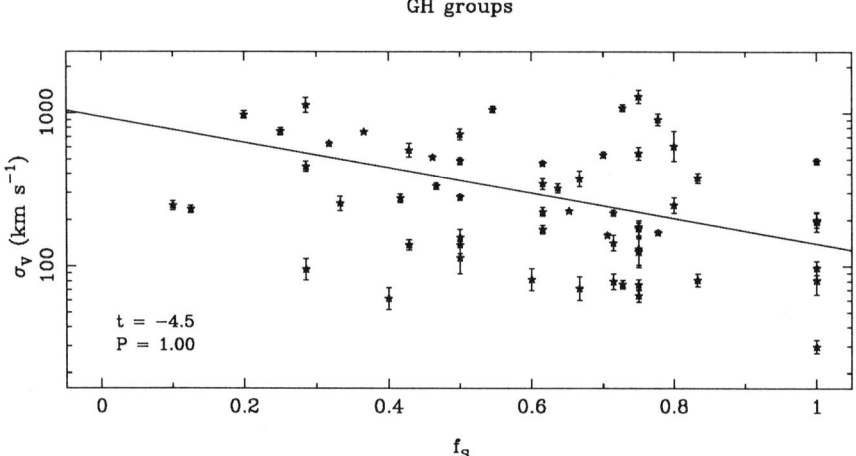

FIGURE 6. Velocity dispersion as a function of the spiral fraction.

high density normal groups and compact groups, which we know are systems of galaxies where the lowest M/L for the highest densities are found.

Related to this last result, we observe the presence of a relation between the velocity dispersion and the spiral content for the GH groups as is illustrated in Figure 6. Relation which is very similar to the one which appears in compact groups (Hickson et al. 1988). This fact again suggests a link between normal and compact groups.

4. SUMMARY AND CONCLUSIONS

Groups are ideal places to perform detailed analysis of properties such as the distribution of mass, or the relation between the dynamics and the morphology. More important, they offer a continuity of characteristics, from high σ_V sparse clouds to dense and low M/L systems, which allow us to test our current ideas about the influence of the environment on the galaxy evolution. The present analysis of the CFA samples of groups indicate that,

- Important statistical errors on the estimation of the mass are detected for the two samples. The main bulk of these errors come from the existence of interlopers.

- The presence of interlopers produces an overestimation of the mass.

- There exist important biases for all the main parameters of the groups for the GH sample. This is not the case for the HG groups, where only a bias on the luminosity appears. Neither M/L nor the crossing time depend on the redshift.

- The main difference between the two samples comes from the distribution of velocity dispersions, which are systematically larger for GH groups, affecting to the crossing time and to the mass estimators. This effect may be due to the selection criteria or by the introduction of real larger systems which begin to appear at higher distances.

- At a good confidence level, the projected and virial mass estimators give similar results for the groups. If there is a difference is in the sense of being the values of the projected mass slightly larger than those of the virial mass.

- The quality of the estimators depend on the shape of the groups, the more elliptical the larger the errors. Departures from spherical symmetry affect to the quality of the mass estimator instead to the estimator itself.

- There is a significant relation between the observed density and M/L which is equivalent to a size−dispersion one. Denser groups tend to show lower M/L.

- There is an evidence of a relation between the spiral fraction and the velocity dispersion.

It seems that there is a connection between normal and compact groups, at least to what respect to the relation of M/L with the density and to the relation of the velocity dispersion with the morphological content. If morphological effects are present in low density systems then part of the problem posed by compact groups may be shared by normal groups.

5. REFERENCES

Barnes, J.E. 1989, *Nature*, **338**, 123.
Geller, M.J. and Huchra, J.P. 1983, *ApJS*, **52**, 61.
Hartwick, F.D.A. 1978, *ApJ*, **219**, 345.
Heisler, J., Tremaine, S., and Bahcall, J.N. 1985, *ApJ*, **298**, 8.
Hickson, P., Kindle, E., and Huchra, J.P. 1988, *ApJ*, **331**, 64.
Huchra, J.P., and Geller, M.J. 1982, *ApJ*, **257**, 423.
Mosteller, F., and Tukey, J.W. 1977, *Data Analysis and Regression*, Addison Wesley, p. 133.
Perea, J., del Olmo, A., and Moles, M. 1990a, *A&A*, **237**, 319.
Perea, J., del Olmo, A., and Moles, M. 1990b, *Ap&SS*, **170**, 347.
Postman, M., and Geller, M.J. 1984, *ApJ*, **281**, 95.
Schechter, P., and Press, W.H. 1976, *ApJ*, **203**, 557.
Thonnat, M. 1989, in *The World of Galaxies*, eds. H.G. Corwin and L. Bottinelli, Springer–Verlag, p. 53.

Groups of Galaxies in CDM Universes

JAMES M. GELB[1]
Massachusetts Institute of Technology, Department of Physics,
Cambridge, MA 02139, USA

ABSTRACT: We identify dark halos from a high resolution $\Omega = 1$ cold dark matter (CDM) simulation. The overmerging of halos forms systems that are too massive to be associated with single galaxies. We convert these systems to groups and clusters assuming a universal mass-to-light ratio. The resulting groups and clusters have a significant affect on spatial clustering and pairwise velocity dispersions, and they pose a serious challenge to the model for any reasonable normalization of the initial power spectrum.

1. INTRODUCTION

We use a 144^3 particle P^{3}M simulation in a 100 Mpc box (Plummer softening of 65 kpc comoving and particle mass of $2.3 \times 10^{10} M_\odot$) to study the evolution of an $\Omega = 1$ (H$_0$ = 50km s Mpc^{-1}) cold dark matter universe. Ben Moore (at this workshop) presented a group analysis from a simulation in a larger volume of space where galaxies were represented by single particles. Although that simulation is useful for testing group identification algorithms, it lacks sufficient spatial and mass resolution to test the CDM model. Our simulation resolves individual halos. However, the lack of gas dynamics, i.e. a dissipative baryonic component, in our simulation is partially responsible for our inability to identify massive systems as groups. Gas dynamical simulations of individual clusters with dark and baryonic matter (Katz and White 1993; Evrard, Summers, and Davis 1992) demonstrate that some galaxies can survive the merging process. Attempts to study gas dynamical effects in large volumes of space suffer from lower resolution yet can offer some insight into the sites for galaxy formation (Cen and Ostriker 1992). We convert massive halos found in our high resolution dark cosmological simulation into groups and clusters assuming a universal mass-to-light ratio in order to understand the group properties and clustering. This has been motivated by studies demonstrating the merging of halos formed at early epochs (e.g., White *et al.* 1987; Gelb 1992). The groups can dominate the statistics and they are essential for understanding if nature can "hide" mass and if there is a normalization of the initial CDM power spectrum matching theory with observations.

We normalize the initial, linear CDM power spectrum, $P(k)$ for comoving wavenumber k, so that the linear, r.m.s. mass fluctuation in spheres of radii $8h^{-1}$ Mpc is σ_8. Larger values of σ_8 correspond to dynamically more evolved

---

[1]PRESENT ADDRESS: NASA/FERMILAB ASTROPHYSICS CENTER, P.O. BOX 500, BATAVIA, IL 60510.

systems:

$$\sigma_8^2 \equiv \int_0^\infty d^3k P(k) W_{\rm TH}^2(kR);$$

$$W_{\rm TH}(kR) = \frac{3}{(kR)^3}(\sin kR - kR\cos kR);$$

$$R = 8h^{-1} \text{ Mpc}.$$

2. TWO-POINT CORRELATIONS

We demonstrate in section 3 that it is necessary to break up massive halos in order to have groups in our simulation. Also, in Gelb (1992), we demonstrated that there are too many halos with $V_{\rm circ} \gtrsim 350 km\ s^{-1}$ and that these systems are too massive to be associated with single galaxies. For these reasons, we break up massive halos into groups, assuming a universal mass-to-light ratio in a procedure outlined below.

The total luminosity in a volume V from the Schechter luminosity function is

$$\mathcal{L}_{\rm total} = V \int_0^\infty \mathcal{L}\Phi(\mathcal{L})d\mathcal{L}\ .$$

The total number of galaxies in a volume V with a luminosity exceeding $\mathcal{L}$ is:

$$N(>\mathcal{L},V) = V \int_\mathcal{L}^\infty \Phi(\mathcal{L})d\mathcal{L}\ .$$

Combining these equations and defining $x \equiv \mathcal{L}/\mathcal{L}_*$, we get the total number of halos exceeding a luminosity $\mathcal{L}$ in a cluster with total light $\mathcal{L}_{\rm total}$:

$$N(>\mathcal{L},\mathcal{L}_{\rm total}) = \frac{\mathcal{L}_{\rm total}}{\mathcal{L}_*} \frac{\int_{\mathcal{L}/\mathcal{L}_*}^\infty x^\alpha e^{-x} dx}{\Gamma(2+\alpha)}\ ,$$

where $\Gamma(2+\alpha) = \int_0^\infty x^{1+\alpha} e^{-x} dx = 1.0456$ for $\alpha = -1.07$.

We take the bound mass of a massive halo (those with $V_{\rm circ} \geq 350 km\ s^{-1}$; see §3) and we divide it by a specified universal value of $M/\mathcal{L}$. This gives us the total luminosity emitted by the cluster: $\mathcal{L}_{\rm total}$. We then add $N(>\mathcal{L},\mathcal{L}_{\rm total})$ halos with luminosity exceeding $\mathcal{L}$.

When we add in halos we need to choose positions and velocities. We do this by randomly sampling positions of particles (velocities are discussed in § 4).

We apply the mass-to-light method and compute the two-point correlation function ξ for halos with $V_{\rm circ} \geq 250 km\ s^{-1}$, see Figure 2. At $\sigma_8 = 0.5$ the enhancement in ξ is nearly sufficient for $M/\mathcal{L} = 50$ but the slope is too steep on small scales. The slope steepens at larger scales for increasing σ_8. The results at $\sigma_8 = 0.5, 0.7,$ and 1.0 have close to the observed correlation length, but the slope at small r is too steep at all σ_8. The no break-up case at $\sigma_8 = 1$ is

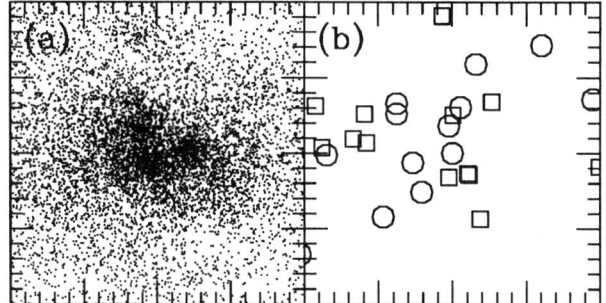

FIGURE 1. A $2.1 \times 10^{14} M_\odot$ halo (a) is split up into a cluster (b). The boxes are 2-D projections of a 2 by 2 by 2 Mpc region. Circles indicate halos with $V_{\rm circ} \geq 250 {\rm km\,s^{-1}}$ and squares indicate halos with $V_{\rm circ} \geq 200 {\rm km\,s^{-1}}$.

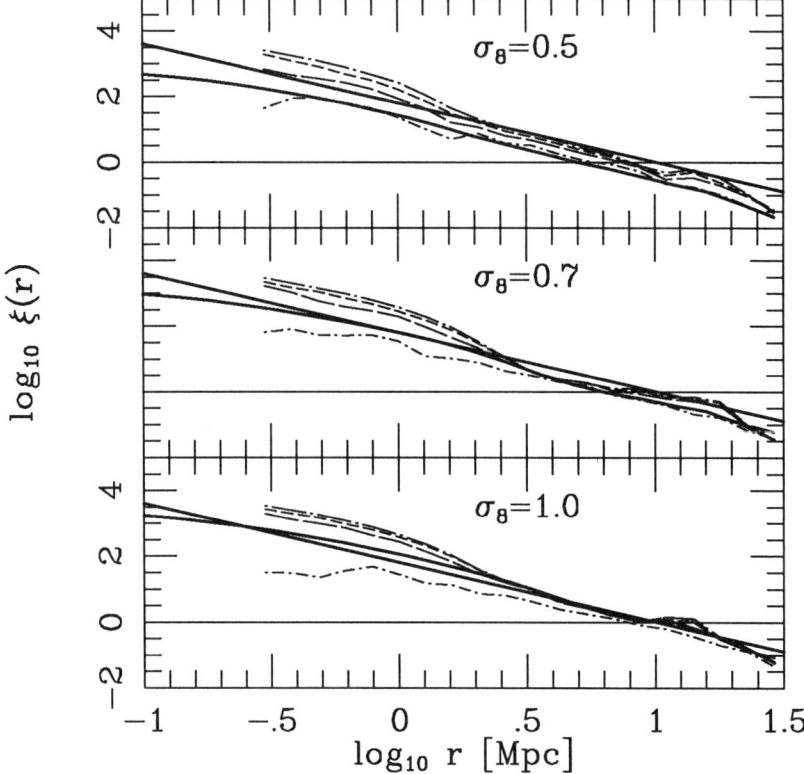

FIGURE 2. Two-point correlations for halos with $V_{\rm circ} \geq 250 {\rm km\,s^{-1}}$. No break-up (dot-short dashed lines); $M/\mathcal{L} = 250$ (long-dashed lines); 125 (short-dashed lines); 50 (dot-long-dashed lines). The straight solid lines are the observed ξ and the curved solid lines are for the mass (all particles).

almost acceptable, but the significant turnover on small scales does not match the observed slope.

White et al. (1987) found a factor of ~ 3 too many halos necessary to enhance ξ at $\sigma_8 = 0.4$ (although they used a 50 Mpc box). We find that the correlation length falls short of the observed value at $\sigma_8 = 0.5$ for $M/\mathcal{L} = 250$ and that breaking up the massive halos also produces factors $\sim 2-3$ too many halos compared with estimates from the Schechter luminosity function.

Before drawing more conclusions, we examine two issues further. The first has to do with mass-to-light ratios for an $\Omega = 1$ universe and the second has to do with the richness of observed groups (§ 3). For $\Omega = 1$, with $h = 1/2$, the mass density is 6.9×10^{10} $M_\odot$ Mpc^{-3}. If we divide this by the blue luminosity density in the universe $9.46 \times 10^7 \mathcal{L}_\odot$ Mpc^{-3} (using $\Phi_* = 1.95 \times 10^{-3}$ Mpc^{-3} for $h = 1/2$) the implied value of $M/\mathcal{L}$ for $\Omega = 1$ is ~ 750. We would be inconsistent with the observed universe using $M/\mathcal{L} \ll 750$ in our $\Omega = 1$ models if the massive halos were characteristic of the universe as a whole. We compute the fraction of the total mass in our (100 Mpc)^3 volume contained in massive halos ($V_{\text{circ}} \geq 350 km\,s^{-1}$). We find the percentage of mass contained in these objects to be 19.2% at $\sigma_8 = 0.5$, 29.9% at $\sigma_8 = 0.7$, and 39.9% at $\sigma_8 = 1.0$. These numbers are large when one recalls that only a few percent of galaxies are in rich groups; see Bahcall (1979) for a review.

3. GROUPS OF GALAXIES

We search for groups and we compare with Ramella, Geller, and Huchra (1989; hereafter RGH) who studied groups of galaxies from the $B(0) \leq 15.5$ CfA-2 redshift survey. For our discussion we convert all relevant quantities to Zwicky magnitudes. We replicate our (100 Mpc)^3 volume using periodic boundary conditions into a (250 Mpc)^3 volume. We then select a wedge corresponding to the CfA-2 sky coverage: right ascension range $8^h \leq \alpha \leq 17^h$; declination range $26.5° \leq \delta < 38.5°$. We assume H$_0$ = 50km s Mpc^{-1} and we impose a distance cut of $R \leq 240$ Mpc in our analysis. We used actual positions rather than redshifts and we imposed an apparent magnitude limit of $B(0) \leq 15.5$. We assume a Tully-Fisher relationship, see Pierce and Tully (1988), relating circular velocity to magnitudes.

We use DENMAX (our local density maxima finder, see Bertschinger and Gelb 1991) to identify all halos with $V_{\text{circ}} \geq V_{\text{circ}}^{\text{MIN}} = 50 km\,s^{-1}$; then we used friends-of-friends (FOF) to identify groups in our wedge after breaking up the massive halos using the mass-to-light method. We determine a FOF linking length, l in Mpc, corresponding to a given overdensity of halos $\delta\rho/\rho$ given by $l^3 = 2/(n\delta\rho/\rho)$ (see, for example, Frenk et al. 1988) where n is the number density of halos with circular velocity exceeding $V_{\text{circ}}^{\text{MIN}}$ from our original (100 Mpc)^3 volume. We use FOF to identify groups of halos after breaking up the massive halos, but prior to imposing an apparent magnitude limit. Typical values of l, for $\delta\rho/\rho = 80$, range from 0.8 Mpc to 1 Mpc for the various assumed values of $M/\mathcal{L}$ and σ_8.

We identify groups with only three or more members to be consistent with RGH. RGH chose a linking distance using a galaxy number density estimated from the observed Schechter luminosity function. However, they varied their linking length with redshift to account for the sparse sampling of galaxies at large redshift. We avoid this difficulty by applying FOF with a fixed linking length prior to applying an apparent magnitude limit. We then apply the apparent magnitude limit to the resulting group catalog in a manner described below.

For field halos, i.e. those that are not in groups with 3 or more members, we simply compute $M_{B(0)}$ using the Tully-Fisher relationship, and we remove those with $B(0) > 15.5$. For the halos in groups we apply the following procedure. If the group member was not created from the break-up of a massive halo, then we eliminate it if $B(0) > 15.5$. For group members that were created from the break-up of a massive halo, we remove all of them and replace them by the number of halos determined using the mass-to-light method with a universal $M/\mathcal{L}$. The lower luminosity limit is computed from $15.5 - M_{B(0)} = 5\log_{10} d + 25.0$, where d is the distance to group centroid in Mpc. Here we do not need to relate circular velocity to luminosity; however, to be consistent with our use of $V_{\rm circ}^{\rm MIN}$, we never allow $\mathcal{L}$ to fall below $\mathcal{L}_{\min}$ determined from $V_{\rm circ}^{\rm MIN}$ using the Tully-Fisher relationship.

We summarize four parameters involved in the identification of groups. 1) We use $\delta\rho/\rho = 80$, the middle value considered by RGH, since we see the same levels of variation with $\delta\rho/\rho$ as reported by RGH and our conclusions do not depend critically on its choice. 2) We use $V_{\rm circ}^{\rm MIN} = 50 km\, s^{-1}$. Our results do not depend sensitively on $V_{\rm circ}^{\rm MIN}$ because the low mass halos quickly fall out of sight. 3) We use $M/\mathcal{L}$=125, 250, and 500. From a list of 36 groups, RGH find a median $M/\mathcal{L}$ of $178h = 89$ for $h = 1/2$. We choose large values of $M/\mathcal{L}$ because, as we will see, even $M/\mathcal{L}$=125 produces groups that are too rich. 4) There is some arbitrariness to the value of $V_{\rm circ}$ above which we break up the massive halos. We use $V_{\rm circ} = 350 km\, s^{-1}$. If we raise this value we get too many isolated massive halos (see Gelb 1992) which are too big to represent individual galaxies. On the other hand, if we lower this value we get even richer groups. However, the numbers of halos added become very small for smaller mass halos and higher mass-to-light ratios.

The results from our simulation are shown in Table 1. We report numbers from RGII for the full $12°$ slice, but we impose a redshift cut of $12000 km\, s^{-1}$. They only study groups with centroids $\leq 12000 km\, s^{-1}$. We report numbers from the simulation for the full $12°$ slice for $R \leq 240$ Mpc. The table shows the number of groups, $N_{\rm grp}$, identified with 3 or more members, with 10 or more members, and with 20 or more members. We also show the number of halos, $N_{\rm gal}$, in the field, i.e. those that are not in groups with 3 or more members. We estimate the CfA-2 field galaxies within 12000 km s^{-1} as follows. The CfA-2 catalog has 1766 galaxies and we estimate from figure 1 in RGH that ≈ 100 galaxies are beyond 12000 km s^{-1}. RGH found 778 galaxies in groups with three or more members and only a hand full of these galaxies are beyond 12000 km s^{-1}. Therefore, the number of field galaxies within 12000 km s^{-1} in the CfA-2 catalog is approximately $1766 - 778 - 100 \sim 900$ galaxies. The last entry in the table, $N_{1/2}$, is a richness statistic defined below.

We can draw several important conclusions from the numbers in the table. If we do not break up the massive halos, then we do not have enough groups and there are no groups with 10 or more members. Therefore we need to break up our massive halos if our simulated universe is to contain groups comparable

TABLE 1. Table 1: Group Statistics in 12° Slice

Data	σ_8	N_{grp} $\geq$ 3 mem.	N_{grp} $\geq$ 10 mem.	N_{grp} $\geq$ 20 mem.	N_{gal} in field	N_{gal} in groups	$N_{1/2}$
CfA-2	N.A.	128	7	2	900	778	6
No Break-Up	0.5	30	0	0	1910	106	< 3
$M/\mathcal{L} = 125$	0.5	136	36	17	1622	1366	19
$M/\mathcal{L} = 250$	0.5	79	19	5	1579	633	11
$M/\mathcal{L} = 500$	0.5	56	6	2	1555	322	6
No Break-Up	0.7	25	0	0	1901	83	< 3
$M/\mathcal{L} = 125$	0.7	197	37	16	1589	1933	19
$M/\mathcal{L} = 250$	0.7	106	17	8	1523	843	14
$M/\mathcal{L} = 500$	0.7	58	8	3	1470	370	9
No Break-Up	1.0	58	0	0	1660	197	< 3
$M/\mathcal{L} = 125$	1.0	237	55	25	1452	2843	22
$M/\mathcal{L} = 250$	1.0	138	27	11	1341	1307	16
$M/\mathcal{L} = 500$	1.0	83	11	5	1300	610	13

to the observed numbers! In all cases we have too many field halos; these are not dominated by faint halos. However, in Gelb (1992) we found that we had the correct number of halos with circular velocities between 150km s^{-1} and 350 km s^{-1}. The reason for this discrepancy is that we have applied only the Tully-Fisher relationship to the halos here rather than a combination of the Tully-Fisher relationship and the Faber-Jackson relationship as we did in Gelb (1992). Applying the Tully-Fisher relationship to elliptical galaxies, which tend to be the most massive halos, makes the halos appear brighter than they really are. On the other hand, most of our group members result from the break-up of massive halos where we do not need to assume a relationship between circular velocity and luminosity. Therefore, we should give more emphasis to the richness of our groups than to the apparent excess of field halos.

We can constrain $M/\mathcal{L} \gtrsim 250$ based on the number of groups with 3 or more members and the total number of halos in all groups with 3 or more members. In most cases, however, we still have too many rich groups with 10 or more members. On the other hand, $M/\mathcal{L} = 500$ appears to do well. The numbers in Table 1 are consistent with the observations at $\sigma_8 = 0.5$, particularly when we consider the fact that the observed number of groups with three or four members are are contaminated by projection effects which can reduce the observed numbers in groups by factors $\gtrsim 30\%$ (see RGH). This lowers the observed numbers in groups and, by definition, raises the observed numbers in the field!

To further quantify the richness of our groups, we compare the cumulative number of halos in groups with the estimates from RGH for the CfA-2 survey.

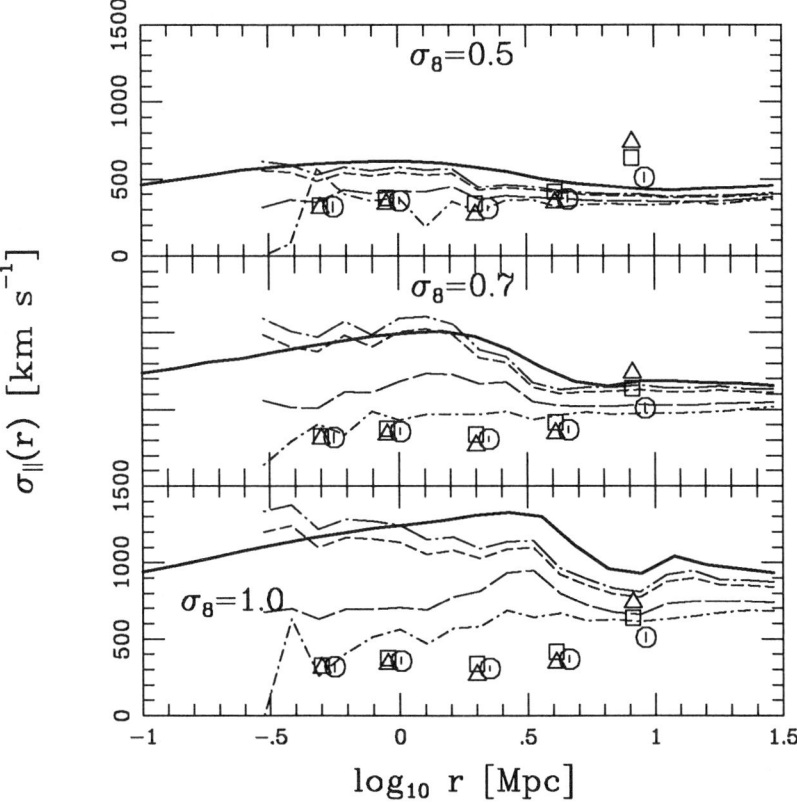

FIGURE 3. Cumulative halo number in groups for various $M/\mathcal{L}$.

The cumulative number of galaxies in groups is defined by RGH as:

$$N_{\text{gal}}(\leq N_{\text{mem}}) \equiv \sum_{N=3}^{N=N_{\text{mem}}} N\, N_g(N)\,,$$

where $N_{\text{gal}}(\leq N_{\text{mem}})$ is the total number of galaxies contained in all groups with three to N_{mem} members and $N_g(N)$ is the number of groups containing N members. The results are shown in Figure 3 which is for a 6° slice and $\delta\rho/\rho = 80$ (we divide the numbers from our 12° slice by two) to compare with RGH (their figure 2).

We clearly see the dramatic shortcoming of the no break-up cases at all epochs. We also find that our groups are too rich compared with RGH; the rise in the cumulative halo number is slower than the results for the CfA-2 survey, indicating that our group members are concentrated in relatively larger groups. A useful statistic is $N_{1/2}$ shown in Table 1. This is the value of N_{mem} where the cumulative number of halos in groups reaches $1/2$ its maximum value. The value of $N_{1/2}$ indicates that we need $M/\mathcal{L} \gtrsim 250$. We can rule out $M/\mathcal{L} = 125$.

We conclude from our studies comparing groups with the CfA-2 survey that we have too many halos in the field. These are dominated by bright halos that have not been broken up. However, this is partly because we have applied the Tully-Fisher relationship to all of our halos rather than a combination of the Tully-Fisher relationship and the Faber-Jackson relationship. Also, the CfA-2 estimates of groups suffer from projection effects; some of the "group" galaxies are actually field galaxies. We also find too many halos in the groups unless $M/\mathcal{L} \gtrsim 250 = 500h$ which is much higher than observed values of $M/\mathcal{L}$. In most cases our groups are too rich with far too many groups containing 10 or more members. We are forced to break up the massive halos; otherwise our catalogs produce far too few groups compared with the observed universe. The case $M/\mathcal{L} = 500$ at $\sigma_8 = 0.5$ gives good agreement, however, with the observed properties of groups. The question is whether or not nature can hide this much mass; this will be an important consideration when we study velocities next.

4. PAIRWISE VELOCITY DISPERSIONS

We now consider constraints on σ_8 from our simulation based on pairwise velocity dispersions ($\sigma_\parallel$) of the resolved halos, see Figure 4. The open symbols in Figure 4 are the observed estimates from the Davis and Peebles (1983) analysis of the CfA $B(0) \leq 14.5$ redshift survey. The different symbols are for different modeling parameters. The details are not important for our purposes; the scatter is small compared with the σ_8 dependence of $\sigma_\parallel$. The results at $r \sim 10$ Mpc are the least accurate because of distortions from peculiar motions.

Based on the no break-up cases in Figure 4, observational data constrains $\sigma_8 \lesssim 0.7$. The case $\sigma_8 = 0.5$ is an excellent match to the observed data. The case $\sigma_8 = 1.0$ is ruled out; the pairwise velocity dispersions are too high by factors ~ 2 for $r \gtrsim 1$ Mpc. Note that this is true even though there is a velocity bias of about a factor of two!

Couchman and Carlberg (1992; hereafter CC), in a slightly lower resolution simulation, found a pairwise velocity dispersion for their halos and their mass which is in reasonable agreement with our result for $\sigma_8 = 1$; however, they only reported results at $r \sim 1$ Mpc. CC did not report pairwise velocity dispersions on larger scales where we find the disparity with the observations to be large. CC also found that their halos are significantly antibiased with respect to the mass (i.e. fall below the mass) on small scales.

We also study pairwise velocity dispersions after breaking up the massive halos. The results are shown in Figure 4 assuming $M/\mathcal{L} = 500$. We randomly sample the positions of the massive halos to assign positions to the added halos. Next we describe how velocities are assigned to the added halos.

We use the 1-dimensional velocity dispersion of each massive halo as the r.m.s. for random numbers (see Gelb 1992); this quantity is labeled $\sigma_1(MH)$ where MH is used to denote the original massive halo. We label the i-th (for $i = x, y, z$) component of the center-of-momentum velocity of the massive halo as $v_i(MH)$. We then compute three gaussian random numbers, r_i, with mean zero and a 1-dimensional standard deviation $\sigma_1(MH)$ for each group member.

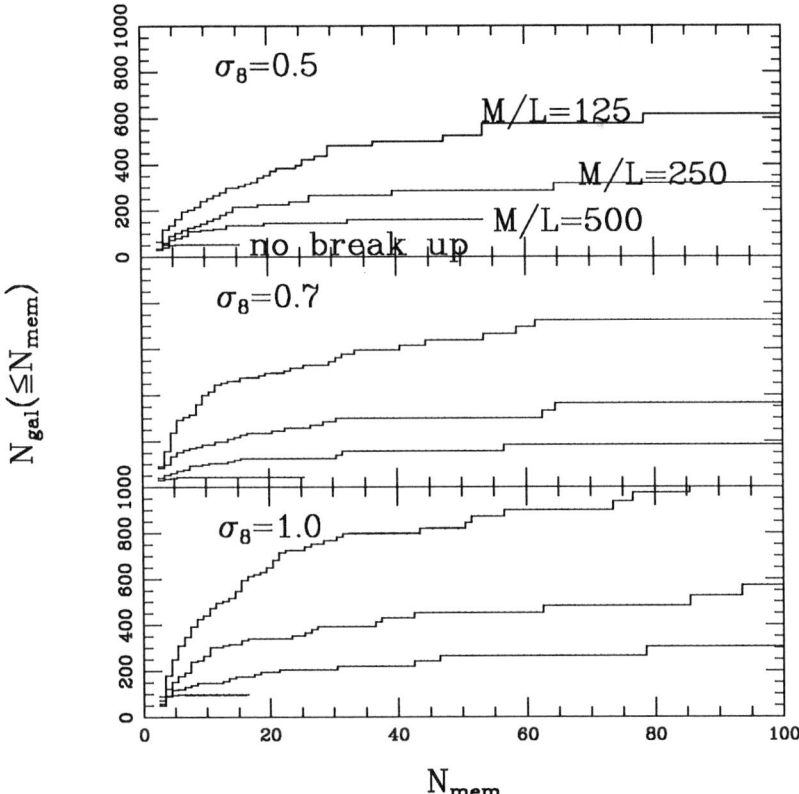

FIGURE 4. Pairwise velocity dispersions for halos with $V_{circ} \geq 150 \mathrm{km\,s^{-1}}$. No break-up (dot-short-dashed lines); mass (solid lines); and $M/\mathcal{L} = 500$ with $\beta = 0.25$ (long-dashed lines), $\beta = 0.8$ (short-dashed lines), and $\beta = 1$ (dot-long-dashed lines).

We define the velocity of the added group member as

$$v_i[\text{group member}] = v_i(\text{MH}) + \beta^{1/2} r_i \;,$$

for some constant $\beta \leq 1$ discussed next.

The quantity β is the ratio of "galaxy temperature" to gas temperature (see Sarazin 1988). The "galaxy temperature" is a measure of the kinetic energy of the galaxies and the gas temperature, which is assumed to be in hydrostatic equilibrium with the group dark matter, is directly related to the velocity dispersion of the group. Observational estimates yield $\beta \sim 0.8$ (Evrard 1990).

Figure 4 shows results with $\beta = 1$, 0.8, and 0.25. These results indicate that the pairwise velocity dispersions are too high at $\sigma_8 = 0.5$, 0.7, and 1.0 if we break up the massive halos. CC did not report the high pairwise velocity dispersions associated with groups. They found that merging decreases the numbers of halos in high dispersion regions, and they referenced Bertschinger and Gelb (1991) where we first discussed why this effect can significantly reduce pairwise velocity dispersions. CC used FOF to identify halos with a prescription for preserving merged systems as distinct halos, but they commented that only $\sim 20\%$ of their "galaxy precursors" survive as distinct "galaxies". This might explain why CC still found antibiasing and lower velocity dispersions. If we demand that the massive halos represent groups and if we use reasonable mass-to-light ratios, then we are forced to add back many halos.

We conclude that even small β cannot save $\sigma_8 \gtrsim 0.7$. The case $\sigma_8 = 0.5$ still has pairwise velocity dispersions that are high compared with the observations for $M/\mathcal{L} = 250$ (not shown) and the model requires $\beta \lesssim 0.25$, which is an extreme value compared with observational estimates (see Sarazin 1988, table 2). The same conclusion holds for $M/\mathcal{L} = 500$ at $\sigma_8 = 0.5$ except that the $\beta = 0.25$ case is a reasonable match to the observed pairwise velocity dispersions.

On the other hand, we find that $M/\mathcal{L} = 500$ at $\sigma_8 = 0.5$ might solve some of the problems with the models. The numbers of halos and group properties are in good agreement with the observations. However, the correlation length ($r_0 \sim 6$ Mpc) falls short of the observed value $r_0 = 10$ Mpc. We have found in this section that the velocities for $M/\mathcal{L} \gtrsim 250$ at $\sigma_8 = 0.5$ are marginally consistent with the observed pairwise velocity dispersions and are in good agreement with the observed pairwise velocity dispersions for $\beta \lesssim 0.25$ and $M/\mathcal{L} = 500$. If CDM is to survive on small scales, nature must conspire to hide a lot of dark matter. Whether or not it can do so is a controversial subject (Peebles 1986).

There is a difficulty in determining the bound mass of halos used in our mass-to-light method. Observationally, mass is inferred dynamically from galaxies. Galaxy tracers do not probe the mass in the outskirts of groups. We are investigating the bound mass of massive halos as a function of local density; imposing a density cut may lower our mass-to-light estimates. In any case, it appears that $\sigma_8 \gtrsim 0.7$ is ruled out in agreement with Davis et al. (1985).

5. CONCLUSIONS

Considerable attention has been paid to massive halos ("godzillas") which might represent groups. Breaking up these massive halos into groups removes the turnover of the two-point correlation function on small scales and it increases the correlation length on larger scales. Unfortunately, the groups do more harm than good unless we assume very high mass-to-light ratios. They significantly increase the number of halos, they give the wrong shape of the two-point correlation function, they significantly increase the pairwise velocity dispersions, and they make groups that are too rich for reasonable mass-to-light ratios.

In agreement with White et al. (1987) we find that we need to restore halos in massive systems to get the required two-point correlation length for $\sigma_8 \sim 0.5$. However, the fact that the model then has a factor ~ 3 too many halos and produces the wrong shape of the two-point correlation function is a serious shortcoming of the model. In agreement with Couchman and Carlberg (1992) we find a velocity bias of a factor ~ 2 for $\sigma_8 = 1$. However, restoring the merged halos in massive systems, which have high velocity dispersions, significantly increases the pairwise velocity dispersions. We can rule out $\sigma_8 \gtrsim 0.7$; even $\sigma_8 = 0.5$ requires $\beta \lesssim 0.25$ which is extremely small compared with observed estimates.

Acknowledgements: This research was conducted using the Cornell National Supercomputer Facility, a resource of the Center for Theory and Simulation in Science and Engineering at Cornell University, which receives major funding from the National Science Foundation and IBM Corporation, with additional support from New York State and members of its Corporate Research Institute. This work is presented in much greater detail in Gelb (1992, Ph.D. thesis) and Gelb and Bertschinger (1992, in preparation). I am indebted to Ed Bertschinger for supervising my Ph.D. research. We thank Neal Katz for useful discussions on groups and we thank Paul Schechter for useful suggestions. This work was supported in part by NSF grant AST90-01762 (at M.I.T.), by the DOE (at Fermilab), and by NASA grant NAGW-2381 (at Fermilab).

6. REFERENCES

Bahcall, N.A. 1979, *ARA&A*, **15**, 505.
Bertschinger, E., and Gelb, J.M. 1991, *Computers in Physics*, **5**, 164.
Cen, R.Y., and Ostriker, J.P. 1992, *ApJ*, **393**, 22.
Couchman, H.M.P., and Carlberg, R. 1992, *ApJ*, **389**, 453.
Davis, M., Efstathiou, G., Frenk, C.S., and White, S.D.M. 1985, *ApJ*, **292**, 371.
Davis, M., and Peebles, P.J.E. 1983, *ApJ*, **267**, 465.
Evrard, A.E. 1990, *ApJ*, **363**, 349.
Evrard, A.E., Summers, F.J., and Davis, M. 1992, preprint.
Frenk, S.F, White, S.D.M., Davis, M., and Efstathiou, G. 1988, *ApJ*, **327**, 507.
Gelb, J.M. 1992, M.I.T. Ph.D. thesis.
Katz, N., and White, S.D.M. 1993, *ApJ*, **412**, 455.
Peebles, P.J.E. 1986, *Nature*, **321**, 27.
Pierce, M.J., and Tully, B. 1988, *ApJ*, **330**, 579.
Ramella, M., Geller, M.J., and Huchra, J.P. 1989, *ApJ*, **344**, 57.
Sarazin, C.L. 1988, *X-ray Emission from Clusters of Galaxies*, Cambridge, Cambridge University Press.
White, S.D.M., Davis, M., Efstathiou, G., and Frenk, C.S. 1987, *Nature*, **330**, 451.

What Determines the Morphological Fractions in Groups and Clusters ?

BRADLEY C. WHITMORE

Space Telescope Science Institute, 3700 San Martin Drive, Baltimore, MD 21218, USA

ABSTRACT: Recent studies indicate that in clusters of galaxies the correlation between morphology and clustercentric radius is more fundamental than the correlation between morphology and local density. In light of this result, we reexamine the morphology-density relation in three other environments: (1) compact groups, (2) loose groups, and (3) the field. In all three cases we find that the relation is very weak or non-existent. It appears that the only location where the morphological fractions are strongly modified is near the centers of clusters.

1. INTRODUCTION

While astronomers can fill volumes with details about the structure and kinematics of galaxies, they cannot explain, with any certainty, why some galaxies are spirals and others are ellipticals. The morphology-density relation discovered by Dressler (1980) should provide a basic clue. Dressler defined the projected density of the ten nearest galaxies on the sky as the "local density." He found that the morphological fractions appear to correlate better with this local density than with the projected distance from the cluster center, a quantity we will call the clustercentric radius. This was somewhat surprising since Melnick and Sargent (1977) had demonstrated the existence of strong morphological gradients as a function of clustercentric distance. It was also surprising since it implied that the local environment around a galaxy did not change; that clusters were not virialized. This led to the current interest in the possible existence of substructure in clusters.

While evidence for substructure has been found in some clusters (e.g., see Fitchett 1988 for a review), Sanromà and Salvador-Solé (1990) have shown that it is not related to the morphological fractions. Simultaneously, Whitmore and Gilmore (1991) found that the clustercentric radius was as good a predictor of the morphological fractions as local density, or possibly slightly better. They found that the fraction of ellipticals is roughly 15% from the edge of a cluster all the way to about 0.5 Mpc from the center, at which point it begins to rise dramatically, reaching values of 60-70% at the very center (Figure 1). In a subsequent paper, Whitmore, Gilmore, and Jones (1992; hereafter WGJ) were able to show that the clustercentric radius was the more fundamental parameter; the local density works fairly well only because it correlates reasonably well with the clustercentric distance. They went on to suggest that the enhancement in the elliptical fraction near the centers of clusters may be due to the destruction of disk or proto-disk galaxies, rather than the formation of a larger fraction of ellipticals. The material from these "failed disk" galaxies would form much of the intracluster medium observed in X-ray observations. This works out quite well both qualitatively (i.e. higher gas-to-stars ratio and lower metalicities in rich clusters) and quantitatively (i.e. correct amount of x-ray gas). In light of these new results, and in keeping with the subject of this workshop, we decided to examine the morphology-density relation in other environments. While it is generally believed that a strong morphology-density relation has been demonstrated in a wide range of environments, especially groups, we shall find that most of these results are due to the inclusion of cluster galaxies in the various samples.

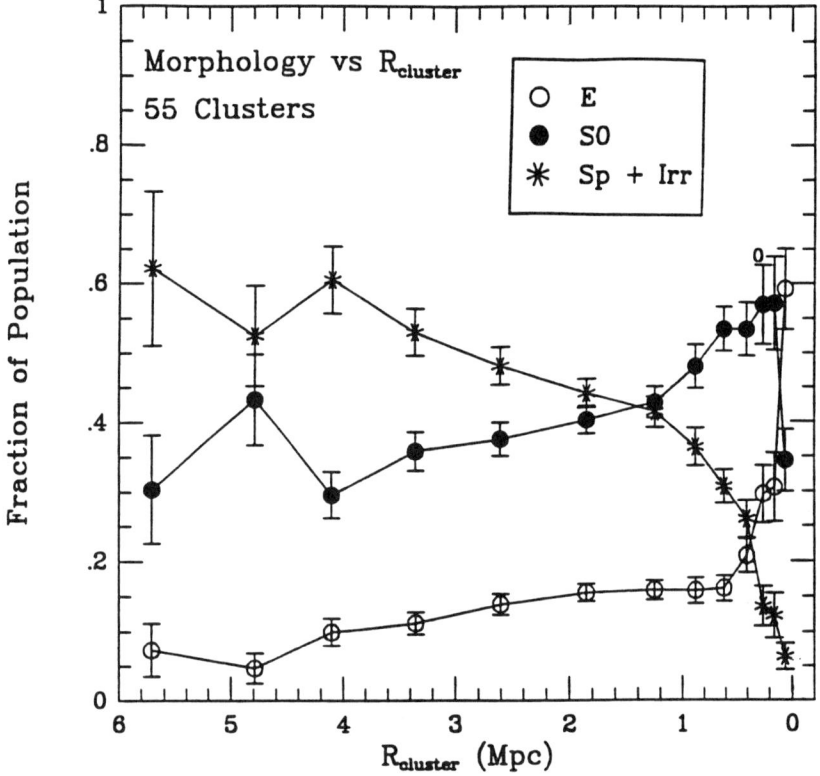

Figure 1. The morphology-clustercentric radius relation for Dressler's (1980) 55 clusters. Note that the elliptical fraction is only high very near the cluster centers. (from WGJ).

2. THE MORPHOLOGY-DENSITY RELATION IN COMPACT GROUPS

Compact groups of galaxies have apparent spatial densities which are as high or higher than even the centers of most clusters. This provides an excellent environment to test for the presence of a morphology-density relation. Using

CCD frames attained on the CFH telescope, Hickson, Kindl, and Huchra (1988) examined the 100 compact groups defined by Hickson (1982). They found a surprisingly weak morphology-density relation (i.e. Spearman coefficient = r = 0.23; 2.4% chance from random distribution). In contrast, they found a much stronger correlation between morphology and the velocity dispersion of the group (r = 0.598; 5.3 x 10^{-7} chance of random) and argued that this indicated that morphology was determined by initial conditions rather than the environment.

Whitmore (1990) found that several properties of the Hickson compact groups correlate strongly with distance, raising the possibility of introducing spurious correlations and repressing real correlations. Whitmore (1992) therefore reexamined this question by removing the distance dependence, but still found a very weak morphology-density relation (linear correlation coefficient = R = 0.14; 18% chance from a random distribution). The morphology-velocity dispersion relation was still the stronger correlation (R = 0.31; 0.6% chance of being random). This is reminiscent of the clusters, where a deep potential well, hence large velocity dispersion, is required to have much effect on the morphology of the galaxies.

Another approach is to compare the population within compact groups with the population in the neighborhoods around the groups. While Rood and Williams (1989) found a weak effect (e.g., elliptical fraction rises from 7 ± 2% in neighborhoods to 15 ± 2% in compact groups), the study by Sulentic (1987) found no significant difference. We conclude that the morphology-density relation in compact groups is either very weak or non-existent.

3. THE MORPHOLOGY-DENSITY RELATION IN LOOSE GROUPS

Probably the strongest evidence for the existence of a morphology-density relation in loose groups comes from the study by Postman and Geller (1984). They studied the problem by making group catalogs of succeeding density from the CfA Redshift Survey (Huchra et. al., 1983) and the Catalog of Nearby Galaxies (Huchra and Geller, 1982) using a friends-of-friends grouping algorithm. They found that at very low-density there was no morphology-density relation: the morphological gradients were constant with the same values as found in the low density regions of clusters. However, for the higher density "groups" they did appear to find a small change in the morphological fractions in the densest few bins (see Figure 2).

A possible problem with this study was the inclusion of cluster galaxies in the sample. While the exact groups used in the study have not been published, we note that in a study by Geller and Huchra (1983) using similar techniques, group #106 was the Virgo cluster with 248 members and group #113 was the Coma cluster with 30 members. Many of the galaxies in the densest "groups" are therefore cluster galaxies. We have repeated the Postman and Geller (1984) approach but with the cluster galaxies excluded. This is done by removing from the original sample any galaxies which find themselves in groups with more than 10 members (when corrected to a fiducial distance appropriate for a galaxy with V = 1000 km s^{-1}, using the D_L scaling described in Huchra and Geller 1983) in the $\Delta \rho/\rho = 100$ catalog (fourth densest catalog). Figure 2 shows that removing the cluster galaxies reduces the morphology-density relation by about 50% (long-dashed lines). The remaining dependence is still suspect, since the densest catalog (right-most bin) shows a strong distance dependence, with most of the ellipticals being in the more distant groups. More details will be published in Whitmore (1993).

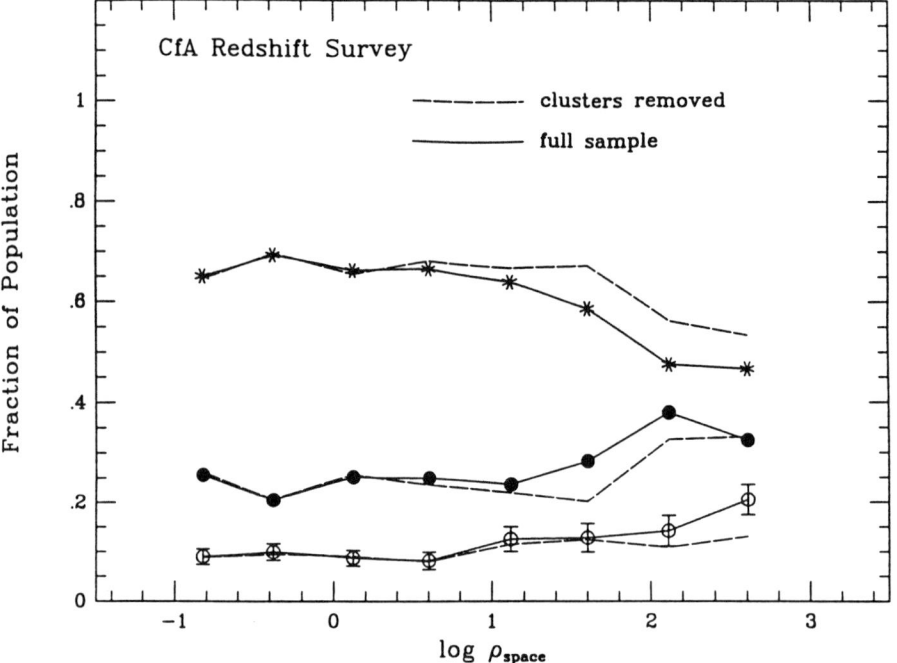

Figure 2. The morphological fractions as a function of space density for the CfA Redshift Survey. The solid lines used the technique described by Postman and Geller (1984). The dotted line used the same technique but with the cluster galaxies removed (see text). The removal of cluster galaxies reduces the morphology-density relation by about 50%. Same symbol definitions as Figure 1.

Perhaps a more straightforward approach would be to simply use a group catalog to examine the morphology-density relation. Figure 3 shows the result for the Geller-Huchra (1983) groups. Groups with more than 10 members have been excluded to reduce the effect of inclusion of cluster galaxies. No hint of a morphology-density relation is seen. Similarly, Maia and da Costa (1990) found no morphology-density relation for the Southern Sky Redshift Survey when treated the same way (their Figure 4), although they do find a weak correlation if they restrict their sample to galaxies with $V = 4000$ km s^{-1} to minimize strong distance dependencies. We conclude that the morphology-density relation is very weak or non-existent in loose groups as well as compact groups.

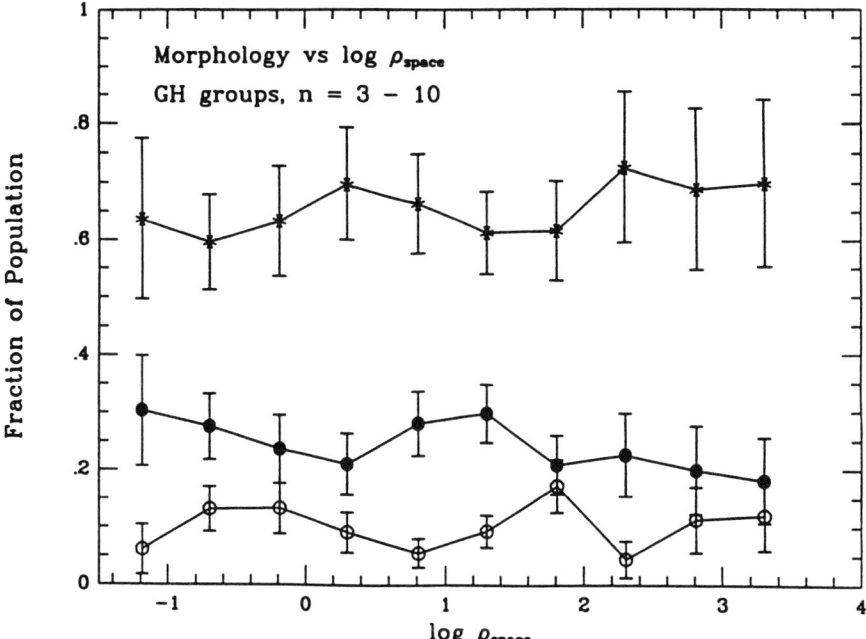

Figure 3. The morphology-density relation for the Geller-Huchra (1983) groups with 10 or fewer galaxies. No correlation is apparent. Same symbol definition as Figure 1.

4. THE MORPHOLOGY-DENSITY RELATION IN THE FIELD

One apparent piece of support for the morphology-density relation is the fact that the distribution of ellipticals on the sky looks different than the distribution of late-type galaxies. However, as stressed by Huchra et al. (1990) for the CfA Redshift Survey, "Outside of the core of the Coma cluster, both early- and late-type galaxies trace essentially the same structure in redshift space." This again implies that the morphological fractions are only strongly affected near the centers of clusters.

We attempted to quantify this subjective impression by using the sample of UGC galaxies in the Pisces-Perseus supercluster, as studied by Giovanelli, Haynes, and Chincarini (1980). At first glance, the distribution of early-type galaxies in this region (E, S0, S0a, compact; their Figure 6b) does appear to look much different than the distribution of late-type galaxies (e.g., Sc, S...; their

Figure 6d). However, this is mainly due to the presence of several Abell clusters in the sample (i.e. A 426, A 347, A 262, A 400) as well as other clusters that fall below Abell's criteria (e.g., Pegasus cluster). If we restrict our attention to regions outside these clusters, the distributions are nearly the same. We defined three high density regions and three low density regions outside of the clusters (e.g., one of the low density regions was 22:50 hr < right ascension < 2:10 hr and 40 deg < declination < 50 deg ; one of the high density regions was 2:00 hr < right ascension < 3:00 hr and 27.5 deg < declination < 37.5 deg). The fraction of early-type galaxies in the high density regions was 17% ± 2% while the fraction of early-type galaxies in the low density regions was 16% ±2%. Hence, there is no evidence for a morphology-density relation in the field for this sample.

Another approach is to calculate the angular correlation function. In fact, one of the first indications that the distribution of early-type galaxies was different than late-type galaxies was the study by Davis and Geller (1976) which computed the angular correlation for the Uppsala Catalog. In most cases no attempt is made to separate the cluster from the non-cluster galaxies. However, in a study of the CfA Redshift Study, Babul and Postman (1990) found that if the Coma cluster is excluded from the sample, the spatial distributions of the late and early-type galaxies are entirely consistent with each other.

Finally, as already discussed in the previous section, Postman and Geller (1984) found no morphology-density relation in the lowest density regions of the CfA Redshift Survey.

5. SUMMARY

The morphology-density relation was originally introduced because of the difficulty of defining the center of a cluster. It spread in popularity because it provided an easy algorithm for quantifying the data and because of a misconception that it was better than the morphology-clustercentric radius relation. However, it now appears that the fundamental correlation is with the clustercentric radius, and much of the evidence for a morphology-density relation in other environments is due to the inclusion of cluster galaxies in these samples. While there are indications that a weak morphology-density relation may exist in some cases, it is important to keep in mind the fact that even if these correlations are real, they generally imply changes of only 5-10% for the morphological fractions, even though the local density is changing by several orders of magnitude. In contrast, the morphological fractions in the cores of clusters change by 40-50% in the space of 0.5 Mpc. The cores of clusters, therefore, represent a much better environment for studying what controls the morphological fractions.

6. REFERENCES

Babul, A., and Postman, M. 1990, *ApJ*, **359**, 280.
Davis, M., and Geller, M. J. 1976, *ApJ*, **208**, 13.
Dressler, A. 1980, *ApJ*, **236**, 351.
Fitchett, M. 1988, in *The Minnesota Lectures on Clusters of Galaxies and Large-Scale Structure*, ed. J.M. Dickey, San Francisco: A.S.P.

Geller, M.J., and Huchra, J.P. 1983, *ApJS*, **52**, 61.
Giovanelli, R., Haynes, M.P., and Chincarini, G.L. 1986, *ApJ*, **300**, 77.
Hickson, P. 1982, *ApJ*, **255**, 382.
Hickson, P., Kindl, E., and Huchra, J.P. 1988, *ApJ*, **331**, 64.
Huchra, J.P., and Geller, M.J. 1982, *ApJ*, **257**, 423.
Huchra, J.P., Geller, M.J., de Lapparent, V., and Corwin, H.G. Jr. 1990, *ApJS*, **72**, 433.
Huchra, J., Davis, M., Latham, D., and Tonry, J. 1983, *ApJS*, **52**, 89.
Huchra, J.P., and Geller, M.J. 1982, *ApJ*, **257**, 423.
Maia, M.A.O., and da Costa, L.N. 1990, *ApJ*, **352**, 457.
Melnick, J., and Sargent, W.L.W. 1977, *ApJ*, **215**, 401.
Postman, M., and Geller, M.J. 1984, *ApJ*, **281**, 95.
Rood, H.J., and Williams, B.A. 1989, *ApJ*, **339**, 772.
Sanromà, M., and Salvador-Solé, E. 1990, *ApJ*, **360**, 16.
Sulentic, J.A. 1987, *ApJ*, **322**, 605.
Whitmore, B.C. 1990, in *Clusters of Galaxies*, eds. W. Oegerle, L. Danly, and M. Fitchett, Cambridge University Press, Cambridge.
Whitmore, B.C. 1992 in *Physics of Nearby Galaxies: Nature or Nurture?*, eds. T.X. Thuan, C. Balkowski, and J.T.T. Van, (Ed. Frontieres, Gif sur Yvette).
Whitmore, B.C. 1993, in preparation.
Whitmore, B.C. and Gilmore, D. 1991, *ApJ*, **367**, 64.
Whitmore, B.C., Gilmore, D., and Jones, C. 1992 *ApJ*, in press (WGJ).

Pairs in Groups and Clusters

JANE C. CHARLTON
Pennsylvania State University, Department of Astronomy,
525 Davey Lab, University Park, PA 16802, USA

BRADLEY C. WHITMORE AND DIANE M. GILMORE
Space Telescope Science Institute,
3700 San Martin Dr., Baltimore, MD 21218, USA

ABSTRACT: Morphological composition of close galaxy pairs is considered in a variety of environments, ranging from "the field" to the centers of Dressler clusters. We find that: 1) Bound pairs with separations up to 100kpc exist in clusters, probably even near the cluster centers; 2) Pairs tend to be elliptical rich, as compared to the expectation based on the local E fraction; 3) There is no evidence for morphological concordance, since mixed pairs are seen as often as pairs where both members are the same type; 4) S0-S0 pairs are not enhanced in clusters, suggesting that ram pressure stripping is not the mechanism for their formation. 5) In low density regions ("the field"), an S0-S pair is more likely, and a S-S less likely, than we would expect from random pairings. We discuss a toy model motivated by the idea that a different type of pair is favored in each different environment. In the context of this toy model, where E's and S0's form by mergers, these morphological tendencies for pairs are related to the density of galaxies and to the local merger timescales.

1. INTRODUCTION

The gross tendency for early type galaxies to occupy higher density environments has been well documented (Dressler 1980). Within galaxy clusters this may be more closely related to the distance from cluster center than to the "local density" determined from the proximity of the ten nearest neighbors (Whitmore, Gilmore, and Jones 1992). The question remains, however, whether galaxies in close pairs show a similar relationship. For example, do members of a close pair tend to contain galaxies of the same type? Do they preferentially contain elliptical galaxies? Do they show the same morphological fractions as nearby unpaired galaxies? These questions are related to the classic theoretical question of "nature vs. nurture", i.e. do galaxies form near other galaxies of the same type, or do certain environments promote similar evolution. For example, in an evolutionary scenario in which mergers lead to the origin of certain morphologies, we might simplistically expect more mixed pairs than in a pure "nature" model.

Mergers of spiral galaxies have often been invoked as a mechanism for the formation of an elliptical galaxy due to the tendency for the merger product to have a smaller angular momentum and a de Vaucouleurs surface brightness profile (Barnes 1988). The details, particularly the behavior of the gas in such a merger, await the implementation of more sophisticated codes such as TREE-SPH (see Barnes and Hernquist 1992). In a previous study emphasizing paired galaxies in low density environments ("the field") of the CfA (Huchra et al. 1983) and SSRS (da Costa et al. 1988) surveys, Charlton and Salpeter (1991) have argued that $\gtrsim 20\%$ of galaxies in these environments are the products of mergers of pairs whose members differ in mass by less than a factor of four. This was based on the observational result of a flat projected separation distribution for bound pairs. The pairs with small separations must merge due to dynamical friction on a timescale of a few orbital periods (much less than a Hubble time for small separations). A merger/replenishment scenario (Ostriker and Turner 1979) can lead to the preservation of a flat distribution and allowed a prediction of the merger rate. The conclusion was a large enough rate that mergers may have created all E's and S0's in these regions. For the numbers to work out correctly we need mergers with mass ratios 4/1 or greater to make S0's, while more equal mass mergers make E's.

In this paper we will describe a study of morphology in paired galaxies from low density environments up to high density groups and clusters. For the low density samples we shall again rely on the CfA and SSRS. Denser regions, including groups and some clusters, will be extracted from the highest density regions of these same catalogs. The Dressler clusters (Dressler 1980) provide our highest density environments, as we proceed from their outskirts in toward their centers. Tabulating the statistics of pairs is particularly difficult in cluster environments. It is not even obvious that bound pairs will exist in clusters, due to the large ($\sim$ 1000km/s) velocity dispersions. Whatever real pairs do exist can easily be confused with the many optical pairs, particularly in the centers of clusters. We attempt to sort this out by establishing the number of optical pairs in a sample with scrambled angular positions, and by considering radial velocity information in the clusters for which it is available.

2. MORPHOLOGY OF PAIRED GIANT GALAXIES IN VARIOUS ENVIRONMENTS

2.1. The Data

A study of pairs in the CfA (Huchra et al. 1983) and SSRS (da Costa et al. 1988) redshift catalogs was described in detail in Charlton and Salpeter (1991). These catalogs, limited to magnitude 14.5, enable the extraction of pairs based on a radial velocity (Δv) criterion, as well as projected separation (r_p). The "density" of the environment of each galaxy was determined by counting the number of neighbors within a truncated cone with $r_p <$ 5Mpc and $\Delta v <$ 600km/s and correcting for the number of galaxies brighter than absolute magnitude -15.7 that will not be seen at that distance in an apparent magnitude limited catalog. (The Hubble constant is taken as 75km/s/Mpc throughout.) If this count results in a number lower than the mean the galaxy is in our "low density half". The galaxies in this half are typically in loose groups of $\leq$ 4 members. The "high density quarter" is similarly defined and contains members of the Virgo cluster, Fornax, and Eridanus, as well as other galaxies in large groups. There will therefore be some overlap between this and our cluster samples.

The Dressler clusters (Dressler 1980) were divided into inner ($<$ 0.33Mpc), moderate (.33 − 1Mpc), and outer ($>$ 1Mpc) regions, where cluster centers were taken as X-ray centers when available and as the cD galaxy otherwise, as in Whitmore, Gilmore, and Jones (1992). We will later make use of radial velocities that are tabulated for about half the galaxies in 14 clusters by Dressler and Shectman (1988) and in 3 clusters by Malumuth et al. (1992).

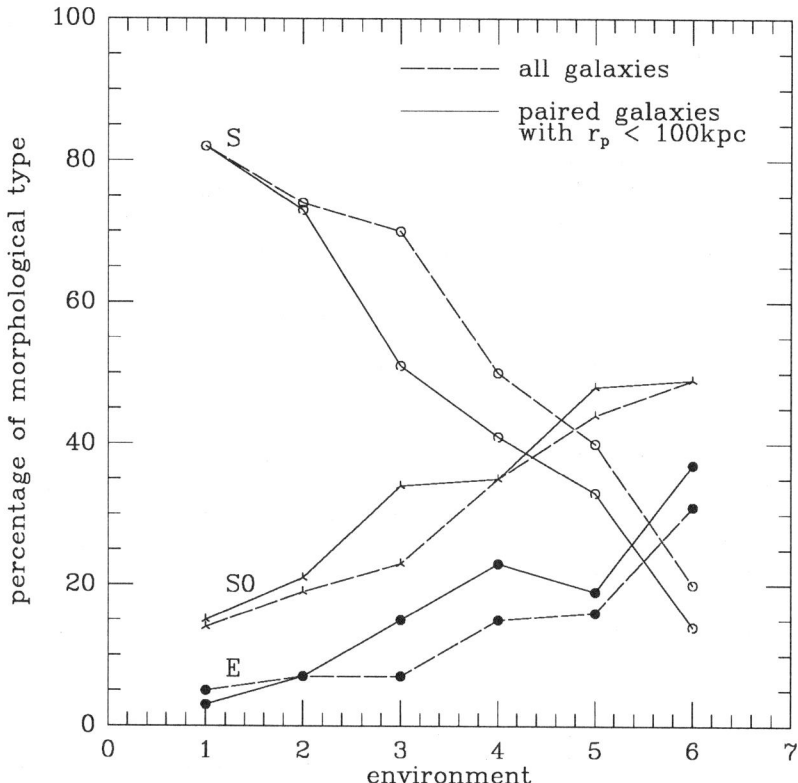

FIGURE 1. Variation of morphology with environment. The following environments are plotted: 1) the low density 1/2 of CfA and SSRS, 2) the high density 1/2 of CfA and SSRS, 3) the high density 1/4 of CfA and SSRS, 4) the outer Dressler cluster regions, 5) the moderate cluster regions, and 6) the inner Dressler cluster regions. The percentages of all galaxies in that environment with a given morphological type are given as solid lines, while dotted lines show the morphological fractions for paired galaxies.

2.2. A Comparison of Various Environments

Fig. 1 shows the morphological composition of all galaxies in six different environments (dotted lines) as compared to paired galaxies with $r_p < 100$kpc in these same environments (solid lines). As we move toward higher density environments the E and S0 fractions increase, while the S fraction decreases. Furthermore, in all environments from the very high density 1/4 up through the various parts of clusters we see that pair members are preferentially E's, while a S is less likely to appear in a pair. In clusters, a paired galaxy is no more or less likely to be an S0 than an unpaired galaxy, however, in the very high density 1/4, which includes groups as well as clusters, pair members are preferentially S0's as well as E's. Caution must be exercised when interpreting these results. Just as the E fraction increases from the outskirts to the inner region of a cluster, it will also increase from the outside to the inside of a bin.

We must address the issue of whether the tendency for pairs to contain E's is merely due to optical pairs toward the inner side of a bin.

2.3. Scrambled Samples: Are Cluster Pairs Real?

Using the real data for the 55 Dressler clusters, we generate a "scrambled" cluster catalog by choosing for each galaxy a random angular position around the cluster center, while keeping its radial distance fixed. Table 1 presents the ratio of E's, S0's, and S's in pairs (with various r_p restrictions) in the unscrambled sample to the number in the scrambled sample, in the inner, moderate, and outer cluster regions. There is clearly an excess of paired galaxies in the real sample for moderate and outer regions of clusters, particularly for very close pairs. Although real paired galaxies exist for all morphological types, an E is more likely to be paired than is an S0 or S. As we would expect, the ratio of the number of pairs in the scrambled sample to the number in the unscrambled sample approaches unity for $r_p > 200$kpc, since these are almost certainly not bound pairs. We have not yet corrected for edge effects, which would only be important for pairs with large r_p in the outer bin. This may produce values slightly in excess of unity for such pairs in Table 1. In the inner bin, the ratio of unity is not a good indication that there are no real pairs. The projected density is so high there, that scrambling is likely to bring any galaxy (paired or not) within 50 or 100kpc of another. Velocity criteria will be a more useful tool in this region of the cluster, and will be described below.

TABLE 1. Ratio of the number of E's, S0's or S's in pairs in the unscrambled sample to the number in the scrambled sample in the inner (0-0.33 Mpc), moderate (0.33-1 Mpc), and outer (> 1 Mpc) regions of the cluster.

Bin	$r_p < 50$ kpc			$50 < r_p < 100$ kpc			$200 < r_p < 400$ kpc		
	E	S0	S	E	S0	S	E	S0	S
0-.33Mpc	1.1	1.2	1.2	1.1	1.1	1.1	0.9	1.0	1.1
.33-1Mpc	2.1	1.6	1.4	1.3	1.3	1.2	1.3	1.0	1.0
> 1Mpc	3.5	2.4	2.4	2.4	1.7	1.7	1.3	1.4	1.3

2.4. What Kind of Pairs are Preferred in Clusters?

We have established that E's are more likely to be members of cluster pairs than are spirals, but we would like to know whether this excess is due primarily to E-E pairs, or whether mixed pairs are common as well. We can estimate the number of "real" pairs of a given type by subtracting the number in the scrambled sample from the number in the unscrambled sample. The number of real and scrambled pairs with $r_p < 100$kpc are compared in Table 2. This table also gives the percentage of real pairs that are E-E, E-S0, E-S, etc., to be compared with the same numbers for random pairings in the same bin (the scrambled sample). We find that, within 1Mpc of the cluster center, an excess of E-E pairs occurs, but E-S0 pairs are also common in the outer regions. This excess is compensated by a lack of S0-S0 and S0-S pairs. The lack of S0-S0 pairings is particularly interesting, because it suggests that these morphological trends are not due to morphological concordance. In fact, taken as a group, the mixed pairs (i.e., E-S0, S0-S, and E-S) are not deficient in number when compared with the concordant pairs (i.e., E-E, S0-S0, and S-S). Again, in the inner bin the numbers are meaningless because we cannot assess the number of "real" pairs.

TABLE 2. For pairs with $r_p < 100$kpc in the three cluster radial bins we give the estimated number of real pairs (unscrambled minus scrambled) and the number in the scrambled sample for the various morphological types. Following the "/" is the percentage of this morphological type.

	E-E	E-S0	E-S	S0-S0	S0-S	S-S
0-.33Mpc "real" scrambled	41/20% 364/15%	56/27% 846/36%	11/5% 188/8%	48/23% 580/24%	41/20% 318/13%	8/4% 80/3%
.33-1Mpc "real" scrambled	49/12% 40/3%	81/20% 224/18%	25/6% 132/11%	107/26% 278/22%	118/29% 393/31%	28/7% 184/15%
> 1Mpc "real" scrambled	37/10% 15/4%	87/23% 42/12%	72/19% 55/16%	31/8% 54/16%	77/21% 107/31%	69/18% 69/20%

2.5. Velocity Criteria for Cluster Pairs

The radial velocity difference for a "bound" galaxy pair in a cluster is likely to be well below the typical velocity difference between cluster members. Thus, any real morphological preference should be stronger for pairs with small Δv, and no longer apparent for those with large Δv. Radial velocities were available for many of the galaxies in 17 of the Dressler clusters. For these clusters, in Table 3 we look at the morphology of $r_p < 100$kpc pairs, separately for $\Delta v < 300$km/s, 300km/s $< \Delta v <$ 600km/s, and 600km/s $< \Delta v <$ 3000km/s. The errors quoted on the percentages of the various types are 1σ. In the inner bin, there is a small tendency for a pair containing an elliptical to have a smaller velocity difference. Although this effect is not statistically very significant, the fraction of pairs that are E-E is substantially larger for $\Delta v < 300$km/s than for $\Delta v > 600$km/s. This is not the case, however, for E-S0 pairs. Thus, toward the center of a cluster it

TABLE 3. For pairs with $r_p < 100$kpc in the 17 Dressler clusters with velocities, the first three columns give the percentage of pair members that are E's, S0's, and S's for three different radial velocity criteria. The last two columns give the percentage of the pairs satisfying that r_p and Δv criterion that are E-E and E-S0. "0" indicates that no pairs of that type were found satisfying the criteria. The three sets of data are for pairs in the inner, moderate, and outer regions of the cluster.

	E	S0	S	E-E	E-S0
Inner Bin					
0-300km/s	50±6%	41±5%	9±2%	31±5%	35±5%
300-600km/s	47±6%	43±5%	10±3%	25±4%	40±5%
600-3000km/s	41±3%	51±4%	8±1%	16±2%	46±9%
Moderate Bin					
0-300km/s	30±6%	48±7%	22±5%	14±4%	22±5%
300-600km/s	14±4%	64±9%	22±6%	0	22±5%
600-3000km/s	23±4%	58±6%	19±3%	6±2%	29±4%
Outer Bin					
0-300km/s	38±9%	34±9%	28±8%	9±5%	34±9%
300-600km/s	23±13%	46±19%	31±5%	0	33±17%
600-3000km/s	11±6%	44±13%	44±13%	0	15±7%

appears that there are bound pairs, and that they are often E-E.

In the outer regions of the cluster, we have already established an excess of E-E and E-S0 pairs by comparing with the scrambled sample. We would like to verify that many of these close pairs have small velocity differences. Unfortunately, velocities are only available for 42% of the galaxies in the outer regions. However, we can still see in Table 3, that E's in pairs are more likely to have smaller velocity differences than the other types. In the outer bin, this is due to E-S0 pairs as well as E-E pairs. (There were no E-E pairs with radial velocity differences exceeding 300km/s.)

2.6. Paired Galaxies in Low Density Regions

In Fig. 1, the low density half of the CfA and SSRS catalogs showed no morphological preference for pair members. Here we will look separately at the six types of pairs rather than just considering whether E's, S0's or S's are favored. In the low density regions, we can be certain that nearly all pairs with $r_p < 100$kpc and $\Delta v < 150$km/s are physically associated. Table 4 gives the fractions of the various types of pairs as compared to the number we would expect on the basis of random pairings. There is a tendency to have more S0-S pairs, and fewer S-S pairs, relative to random pairings. In fact, this same trend applies up to

TABLE 4. For low density half pairs with $r_p < 100\text{kpc}$ and $\Delta v < 150\text{km/s}$, this table gives the percentage of various types, and the percentage expected for random pairings in these regions.

	E-E	E-S0	E-S	S0-S0	S0-S	S-S
"real"	0%	2%	5%	5%	22%	66%
random	0%	1%	4%	2%	13%	80%

separations as large as 400kpc, another indication that dark matter halos are extensive enough to bind pairs at those separations.

3. SUMMARY AND DISCUSSION

First, let us summarize the various morphological tendencies that have been extracted from the data: 1) Comparison of a cluster sample, with scrambled radial positions, to the real sample demonstrated that bound pairs do exist, at least in the outskirts of clusters. Furthermore, the predominance of close E-E pairs with small Δv indicates that such pairs may even remain bound in the central regions of clusters; 2) The various cluster morphological tendencies persist to separations of $\sim$ 100kpc, suggesting that dark matter halos extend to roughly this distance from galaxies in clusters; 3) There is no evidence for morphological concordance. Although E-E pairs are more likely than expected from random pairings in that region of the cluster, S0-S0 and S-S pairs are less likely. 4) In low density regions ("the field"), a larger fraction of S0's are paired than S's. Both S0-S0 and S0-S pairs occur more frequently than expected from random pairings. This trend still holds for pairs with separations as large as 400kpc.

Our study focused on pairs in which both members are giant galaxies, however there may be certain parallels or clues in studies of dwarf companions. The ratio of dwarfs to giants increases by a factor of twenty from isolated environments to clusters (Vader and Sandage 1991; Ferguson and Sandage 1991). However, the number of very close dwarf companions (projected separations < 80kpc) is a factor of three smaller in clusters than in the field (Ferguson 1992). This suggests that tidal stripping can separate dwarfs from cluster giants, with the dwarfs remaining in the cluster. Ferguson (1992) also found that early type galaxies (E's and S0's) in clusters have twice as many companions as spirals within 80kpc. This is similar to our results both in the scale on which pairs can be bound, and in the tendency for E's to have more smaller, as well as more similar mass, companions. Ferguson's (1992) proposed that ellipticals, as the most massive members of the cluster, would be least susceptible to tidal stripping, and could thus retain more dwarf companions. Similar explanations should be explored for more equal mass pairs.

It is significant that there is evidence for bound pairs in clusters, particularly in the centers. It seems that the paired galaxies must have 100kpc dark matter halos to remain bound in a rather hostile environment. However, dynamical friction of overlapping halos will promote the eventual merger of the pairs. It remains to be determined whether the time scale for merger would be short enough to deplete the cluster of systems that were paired from the time of cluster formation. Alternatively, they could be the result of collisions, although the large cluster velocity dispersion would prevent capture in some cases.

4. A TOY MODEL FOR THE ORIGIN OF MORPHOLOGY

Let us consider a toy model in which all giant galaxies began as spirals. Mergers between roughly equal mass galaxies make ellipticals, while mergers between unequal mass galaxies (eg. 4/1 or greater mass ratio) make S0's. For a pair of galaxies separated by 1Mpc in the low density region, a merger occurs in roughly one Hubble time. If the separation is smaller, as it would be for some regions with larger local densities, the galaxies will merge on a shorter timescale. Studies of the correlation functions of galaxies tell us that if a galaxy is close to another there is more likely to be a third nearby. Thus, we suggest that in relatively low density regions, there will be many S-S0 pairs that resulted from an initial group of three galaxies. This is a region for which the density is not so large as to lead to a merger of all three on the scale of a Hubble time.

The formation of an elliptical would be relatively rare in a low density environment, because the mass function will most likely lead to a massive galaxy with a few lower mass companions. In a region with a dozen galaxies, however, there may be a couple of nearly equal mass galaxies that would have time to merge by the present. This process would still most likely begin with the merger of some companions of giant galaxies, leading to a group of S's and S0's. In a moderate density region, we might end up with a number of S0-S0 pairs. However, in order to make a pair containing an elliptical we need at least three more equal mass galaxies. If we begin with three spirals surrounded by a number of smaller companions, Fig. 2a illustrates the possible progression of events.

The largest density fluctuations will produce the most galaxies, and will ultimately end up as the centers of clusters. Here, much of the merging action may occur rather early on in the evolution of the grouping, before stripping reduces the effects of dynamical friction. This is based on the idea that a cluster is formed by the assembly of many groups. However, the lack of a strong morphology-density relation in groups (Whitmore, this volume) may be a problem for this interpretation. The centers of clusters could once have been groups of many equal mass galaxies with numerous companions as in Fig. 2b. The progression may be halted before reaching stage six. This simple model would explain why S-S pairs are rare everywhere, S0 rich pairs are found in low density environments, and E rich pairs are found in denser environments.

There are alternatives to this simple toy model. For example, rather than assuming that two ellipticals in a cluster are more likely to from a pair than is some other morphological mix, we can consider the possibility that the lifetime of a pair is dependent on the morphological make-up. The mass of a galaxy could depend on the type, with fewer tidal disruptions of the more massive

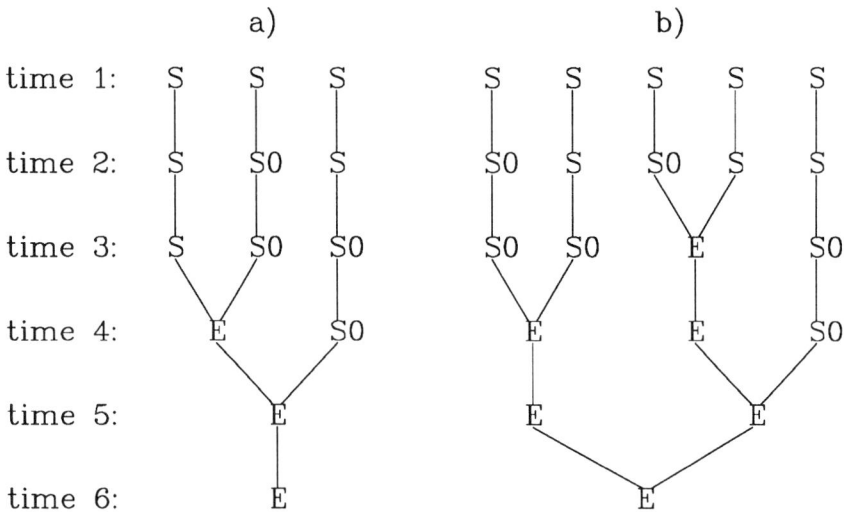

FIGURE 2. A toy model for evolution of morphological types in a merger scenario.

systems. Another alternative is based on the idea that a disk can only form in a protected environment. If there is a nearby galaxy the disk might never form in an extended proto-spiral, thus a spiral would only occur in a pair if captured later. It is, however, difficult to understand how an E-E pair formed early on by such failed spirals would persist for so long. If the pair members were close enough together to prevent the formation of a disk, we might expect the merger timescale to be short.

Acknowledgement: This work was supported by NSF grant AST88-22297 at Steward Observatory of the University of Arizona.

5. REFERENCES

Barnes, J.E. 1988, *ApJ*, **331**, 669.
Barnes, J.E., and Hernquist, L. 1992, *ARA&A*, **30**, 705.
Charlton, J.C., and Salpeter, E.E. 1991, *ApJ*, **375**, 517.
da Costa, L.N., *et al.* 1988, *ApJ*, **327**, 544.
Dressler, A. 1980, *ApJ*, **236**, 351.
Dressler, A., and Shechtman, S.A. 1987, *AJ*, **95**, 284.
Ferguson, H.C. 1992, *MNRAS*, **255**, 389.
Ferguson, H.C., and Sandage, A. 1991, *AJ*, **101**, 765.
Huchra, J.P., Davis, M., Latham, D., and Tonry, J. 1983, *ApJS*, **52**, 89.
Malumuth, E.M., Kriss, G.A., Dixon, W.V., Ferguson, H.C., and Ritchie, C. 1992, *AJ*, **104**, 495.
Ostriker, J.P., and Turner, E.L. 1979, *ApJ*, **234**, 785
Vader, J.P., and Sandage, A. 1991, *ApJ*, **379**, L1.
Whitmore, B.C., Gilmore, D.M., and Jones, C. 1992, *ApJ*, in press.

Groups in the Las Campanas Deep Redshift Survey: A First Look

DOUGLAS L. TUCKER
Department of Astronomy, Yale University, New Haven, CT 06511, USA

ABSTRACT: Presented are findings from a preliminary group catalogue derived from the $\delta_{1950} = -6°$ slice of the Las Campanas Deep Redshift Survey.

1. INTRODUCTION

In the past decade, redshift surveys have been an efficient and popular means of mapping the large-scale spatial distribution of galaxies in the nearby Universe (cf. Giovanelli and Haynes 1992 for an excellent review of recent redshift surveys). Unfortunately, up till now, no survey has been large enough to sample a fair volume of the Universe. These surveys have always contained structures – walls, sheets, filaments, or voids – with dimensions on at least the order of the surveys themselves. The objective of the Las Campanas Deep Redshift Survey (LCDRS) is to encompass a fair volume of the Universe.

The LCDRS will eventually contain ≈25,000 galaxies and cover ≈1,000 square degrees in the North and South Galactic Caps out to a redshift of z = 0.2. The survey pattern in each cap is that of alternating 1.5°-thick slices in declination. The survey is magnitude limited, including those galaxies having isophotal Kron-Cousins red magnitudes of $16.0 < R_{K-C} < 17.3$. The galaxies are selected from a photometric catalogue produced via CCD drift scans on the Las Campanas Swope 1-meter; the spectra are obtained with a plug-plate multiobject spectrograph at the Las Campanas DuPont 2.5-meter. At present, ≈10,000 galaxy spectra have been obtained, of which ≈5,000 have been reduced to yield redshifts. Details concerning the technical aspects and early results of the LCDRS can be found in Kirshner et al. 1991, Shectman et al. 1992, and Shectman et al. 1993.

At present, the most complete slice lies at dec1950 = -6d. A plot of the galaxy positions is shown in Figure 1, in which, for comparison, one of the CfA Slices is reproduced to scale (Geller and Huchra 1989). The two darkened sectors are regions toward which the survey has yet to be extended. Note that the same types of structure in the CfA Survey, e.g. voids and walls, are apparent in the LCDRS. Note also that the largest sharp structures seem to be no larger than 100 h-1 Mpc (h = Ho/100 km/s), much smaller than the dimensions of the survey, making a qualitative argument for the fairness of the sample.

2. THE GROUP CATALOG

A group catalogue for the $-6°$ Slice was derived in the following manner:

1. Associations of galaxies were found via a "friends-of-friends" percolation algorithm as described in Huchra and Geller (1982) and modified for use

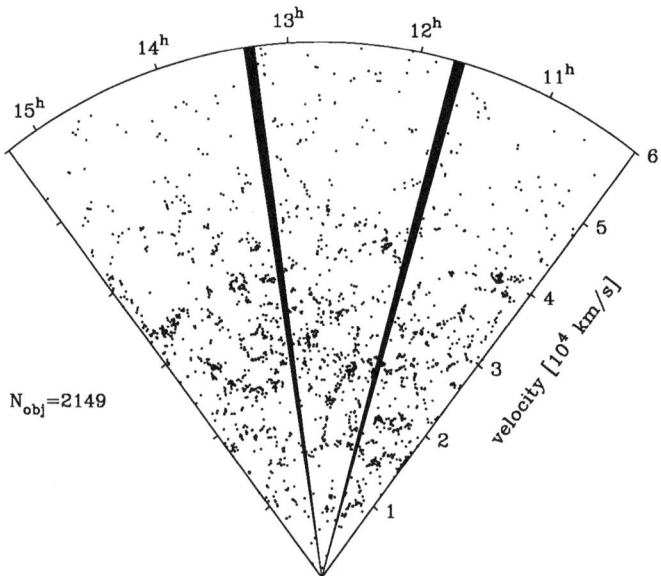

FIGURE 1. The LCDRS −6° Slice. The two shaded sectors are regions toward which the survey has yet to be extended. For comparison, a slice of the CfA survey, reproduced to scale, is included to the side (from Geller and Huchra 1989).

with comoving distances. Associations are defined to contain at least three members.

2. Groups were then identified as associations which

   (a) are contained within number density enhancements above a certain value above the global mean, and

   (b) have crossing times less than a Hubble time (H_o^{-1}).

For the −6° Slice, all but 4 of the 166 associations within a density contrast contour of $\delta\rho/\rho = 40$ met this definition of a group. Unless noted otherwise, only those 162 associations classified as groups will be discussed for the remainder of the paper. In the percolation algorithm, a velocity-linking parameter of $V_L = 700$ km/s was used at a fiducial velocity $V_{fid} = 30,000$ km/s. The galaxy luminosity function was parameterized according to a Schechter function with $\Phi^* = 0.015$ galaxies/magnitude/Mpc^3, $M^* = -20.0$, and $\alpha = -0.65$, assuming $H_0 = 100$ km/s/Mpc.

3. RESULTS

Figure 2 shows the $\delta\rho/\rho = 40$ group members from the −6° Slice; Figure 3 shows the group centers. Not unexpectedly, the group members and the group centers tend to delineate the large-scale structures seen in Figure 1. Groups tend to lie on the walls surrounding voids. Due to poor sampling of galaxies at large distances, however, the distribution of groups is strongly incomplete beyond a redshift of $z \sim 0.15$.

Group properties are calculated following the prescription of Ramella, Geller, and Huchra (1989; henceforth RGH1989), partially modified to take into account comoving distances. The harmonic radius is calculated as

$$R_h = \pi D_c \sin\left\{\frac{1}{2}\left[\frac{N_{obs}(N_{obs}-1)}{2}\left(\sum_i \sum_{j>i}\Theta_{ij}^{-1}\right)^{-1}\right]\right\}, \quad (1)$$

where Θ_{ij} is the angular separation of group members i and j, N_{obs} is the number of observed group members, and D_c is the comoving distance

$$D_c = \frac{c}{H_o q_0^2 (1+z)}\left[q_0 z + (q_0 - 1)\sqrt{1+2q_0 z} - 1\right] \quad (2)$$

(Mattig 1958, Sandage 1988). A value of $q_0 = 0.1$ was used for the deceleration parameter. The mean pairwise separation is

$$R_p = \frac{8D_c}{\pi}\sin\left[\frac{1}{N_{obs}(N_{obs}-1)}\sum_i\sum_{j>i}\Theta_{ij}\right], \quad (3)$$

and the virial mass is

$$M = \frac{6\sigma_{los}^2 R_h}{G}, \quad (4)$$

where σ_{los} is the group's line-of-sight velocity dispersion. The virial crossing time is

$$t_{cross} = \frac{3}{5^{3/2}}\frac{R_h}{\sigma_{los}}. \quad (5)$$

Table 1 contains values for some of the physical properties of the observed $\delta\rho/\rho = 40$ groups, along with values from RGH1989 from their study of groups in the CfA $26.5° < \delta_{1950} < 38.5°$ slice (which is pictured in Figure 1). Since

$$M/L_{B(0)} \sim 1.1 M/L_R, \quad (6)$$

we can estimate that the median $M/L_{B(0)} \sim 300 h M_\odot/L_\odot$ for LCDRS groups.

The differences between the LCDRS and the RGH1989 catalogues are not as glaring as one might suppose at first glance. From equations 4, 5, and 6, we see that the crossing time, virial mass, and mass-to-light ratios depend linearly on R_h. The factor of two difference between the derived LCDRS and RGH1989 group masses and crossing times is due to the factor of two difference between the two catalogues' values for R_h. The primary reason for this discrepancy (and for the discrepancy in the two catalogues' values for R_p) is that RGH1989 defines groups as associations of galaxies within density contours which are ~ 3 times as dense as those used in the LCDRS catalogue. Thus, the RGH1989 groups are naturally smaller in volume than the LCDRS groups. In other words, LCDRS groups are looser groups than those of RGH1989. Most of any remaining

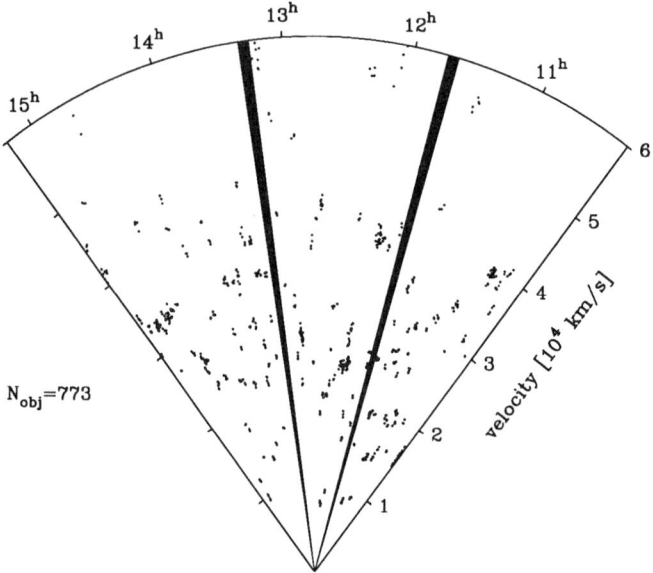

FIGURE 2. Group members in the −6° Slice.

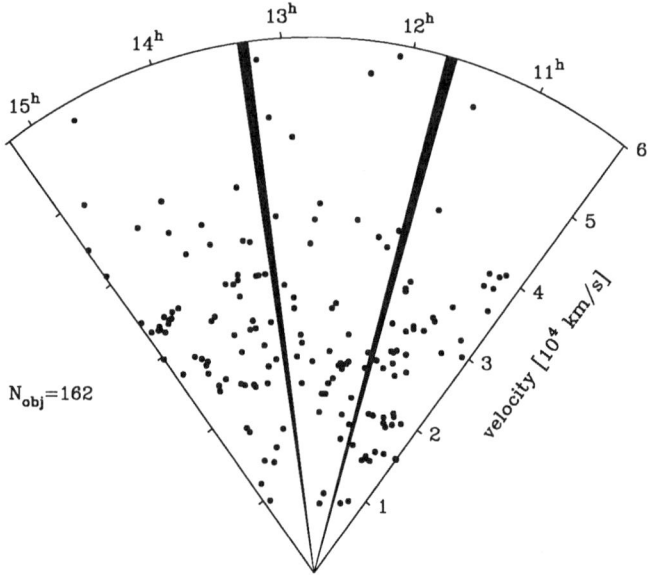

FIGURE 3. Group positions in the −6° Slice.

TABLE 1. Group Properties

Property	LCDRS	RGH1989 (all groups)	RGH1989 (rich groups)*
% of galaxies within groups	36%	44%	13%
median σ_{los} [km/s]	210	209	228
median t_{cross}/t_{Hubble}	0.14	0.06	0.06
median R_h [h^{-1} Mpc]	1.10	0.51	0.52
median R_p [h^{-1} Mpc]	1.26	0.67	0.69
median M [10^{13} h^{-1} $M_\odot$]	5.75	2.00	3.89
median L [10^{11} h^{-2} $L_\odot$]	1.70 (RK-C)	—	—
median M/L [h $M_\odot/L_\odot$]	272 (RK-C)	186 (B(0))	175 (B(0))

*RGH1989 rich groups are those groups containing 5 or more members.

discrepancy in the values for R_h, R_p, and related quantities can be attributed to differences in the sample volumes from which the two catalogues were derived.

In closing, it would be interesting to know how many of the LCDRS groups are virialized. Following Gott and Turner (1977), it can be estimated that the time for a uniform sphere to undergo complete virialization is

$$t_{virial} \sim 3\pi t_{cross}. \qquad (7)$$

Hence, all groups with t_{cross} less than $\sim 0.11 t_{Hubble}$ should have had enough time to virialize completely within the age of the Universe. The distribution of crossing times for all 166 LCDRS associations is displayed in Figure 4, from which we can deduce that somewhat less than half of the groups could have undergone complete virialization.

4. THE FUTURE

This has been just a preliminary analysis of groups within the LCDRS. Future plans include continued testing and improvement of the group-finding algorithm, and cataloguing groups in further slices of the LCDRS. In addition, it is hoped that a study of environmental effects on galaxy properties – for example, tests for luminosity-density and color-density relationships – can be performed with the percolation algorithm. Research into group-group clustering properties is also under consideration.

Acknowledgments Many thanks to Gus Oemler, Ben Moore, and Carlos Frenk, for their many helpful suggestions. This work has been supported by NSF grant AST 89-21662.

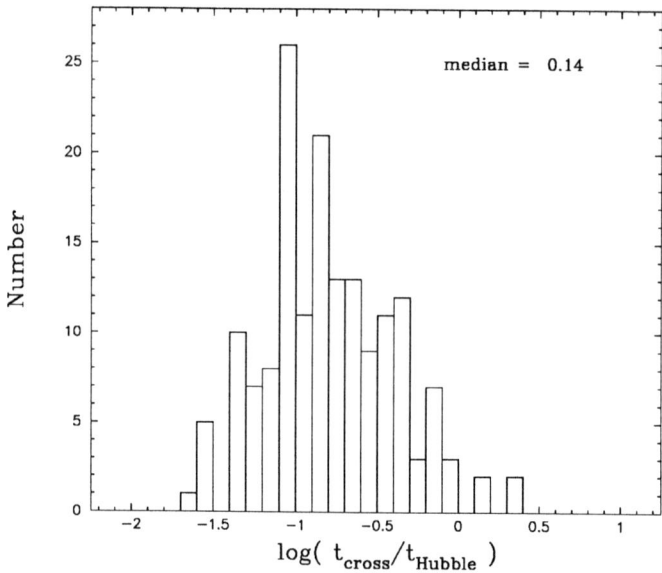

FIGURE 4. Distribution of crossing times for all galaxy associations in the $-6°$ Slice. All associations with crossing times t_{cross} less than a Hubble time t_{Hubble} are considered groups. Those groups with $t_{cross} < 0.11 t_{Hubble}$ have had time to undergo complete virialization within the lifetime of the Universe.

5. REFERENCES

Geller, M.J., and Huchra, J.P. 1989, *Science*, **246**, 897.
Giovanelli, R., and Haynes, M. 1992, *ARA&A*, **29**, 499.
Gott, J.R. III, and Turner, E.L. 1977, *ApJ*, **213**, 309.
Huchra, J.P., and Geller, M.J. 1982, *ApJ*, **257**, 423.
Kirshner, R.P., Oemler, A., Schechter, P.L., Shectman, S.A., and Tucker, D.L. 1991, in *Physical Cosmology*, proceedings of the the Second Rencontre de Blois, (Gif-sur-Yvette Cedex, France: Editions Frontieres), p. 594.
Mattig, W. 1958, *Astron.Nachr.*, **284**, 109.
Ramella, M., Geller, M.J., and Huchra, J.P. 1989, *ApJ*, **344**, 57.
Sandage, A. 1988, *ARA&A*, **26**, 561.
Shectman, S.A., Schechter, P.L., Oemler, A.A., Jr., Tucker, D.L., Kirshner, R.P., and Lin, H. 1992, preprint.
Shectman, S.A., Schechter, P.L., Oemler, A.A., Jr., Tucker, D.L., Kirshner, R.P., and Lin, H. 1993, in preparation.

Comments on the Distribution of Mass Within Dark Matter Halos

DENNIS ZARITSKY[1]
The Observatories of the Carnegie Institution,
813 Santa Barbara St., Pasadena, CA 91101, USA

ABSTRACT: I discuss recent studies of the distribution of mass within the dark matter halos of spiral galaxies. The dynamics of satellite galaxies of both our galaxy and other galaxies indicate that the dark matter halos extend at least 200 kpc and contain at least $10^{12} M_\odot$. Although my focus is on the halos of single galaxies, many of the same issues are relevant to the study of groups of galaxies.

1. INTRODUCTION

The distribution of dark matter in spiral galaxies beyond radii probed by neutral hydrogen rotation curves is difficult to measure. Despite the hope that binary galaxies could be used to probe the outer halo, no consensus has developed from studies of the dynamics of binary galaxies.

To progress beyond this impasse and develop a new approach, it may help to consider the ideal system for study - an isolated spiral galaxy surrounded by an endless spherical cloud of orbiting massless test particles. Of course, each element of this ideal system is highly unrealistic. However, a sample of objects that best approximates this ideal will provide the greatest opportunity for progress. First, although a truly isolated galaxy does not exist, isolation criteria, which minimize the possible effects of nearby massive neighbors are essential. Second, although massless probes are an idealization, low mass probes should be used to minimize the distortion of the primary's gravitational potential by the secondary. Satellite galaxies are an obvious choice because they are significantly less massive than the primary and yet are bright enough to observe fairly easily. Third, because the typical spiral galaxy has only one or two relatively bright companions (Lorrimer et al. 1992), an ensemble of primaries must be used to reduce statistical uncertainties. Finally, although arbitrarily large separations cannot be probed, the sample must contain companions at large separations from the primary. To strongly constrain a set of models out to a particular radial distance from the primary, it is necessary to have data for a significant number of companions beyond the radius of interest. As I will discuss, careful consideration of each of these components of the ideal system leads to better sample definition and a clearer understanding of the issues at hand. Over the past few years, samples of satellites of both our galaxy and other galaxies have become available.

In this article, I discuss the necessary improvements in both sample selection and analysis technique that lead to unambiguous results regarding the distribution of dark matter within a few hundred kpc around the luminous components of spiral galaxies. Many of the points I shall stress, such as those regarding

---

[1] Hubble Fellow.

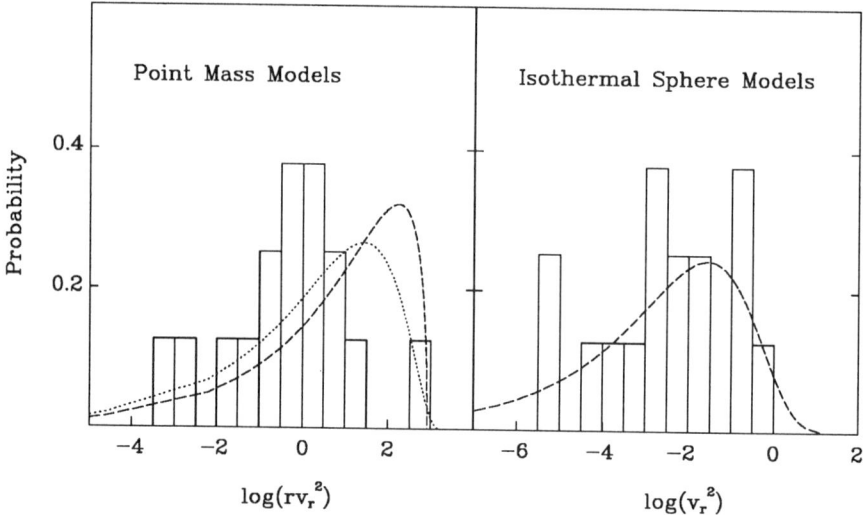

FIGURE 1. Comparison of Galactic Halo Models and Remote Satellite Data. In the left panel the histogram for observed satellites is compared to predicted distributions for point mass models (radial orbits - dashed line; orbits with isotropically distributed velocities - solid line) with best fit total mass. In the right panel the analogous comparison is made to isothermal sphere models. Note that for these models the radial and isotropic velocity models predict the same distribution (from Zaritsky et al. 1989).

proper treatment of contamination by background galaxies and the connection between the small-scale dynamics and the development of structure on larger scales, are equally valid when dealing with groups of galaxies. In §2 I discuss results obtained from the study of companions of our galaxy. In §3 I continue by discussing some results obtained from the study of companions of other galaxies. Primarily, I want to communicate to the reader that studies of our galaxy and other galaxies are now in accord on the size and mass of galactic halos out to at least 200 kpc.

2. COMPANIONS OF OUR GALAXY

There are at least two reasons why the study of companions of our galaxy is simpler than the study of companions of other galaxies. First, our galaxy has the richest known system of satellites because we can observe intrinsically fainter companions. Second, we know the true distance, instead of the projected separation, between the companions and the center of the potential, On the other hand, because it is just one system of satellites, it is difficult, and at some point becomes impossible, to increase the number of available "test particles". It may be impossible to obtain a definitive result.

There have been several studies of the dynamics of satellites of the Galaxy since that by Hartwick and Sargent (1978). Recently, Little and Tremaine (1987) presented an analysis of a preferred subsample of Galactic satellites from the Olszewski, Peterson, and Aaronson sample (1986). They derived 90% confidence limits on the mass of the Galaxy that were surprisingly low ($M < 5.0 \times 10^{11} M_\odot$ for isotropic orbits) given the standard, although at the time weakly supported hypothesis, that galaxy halos are large and massive i.e. $M/L \gtrsim 100$, size > 100 kpc, $M > 10^{12} M_\odot$). Since 1987, the one major weakness of the sample of Galactic satellites has been addressed. The velocities of the most distant systems, which had the poorest velocity measurements and so were excluded in the preferred subsample, were remeasured at higher precision (Zaritsky et al. 1989). An analysis of the complete sample demonstrated that there were systematic difficulties with the Little and Tremaine results. The inclusion of the five newly measured systems increased the estimated mass above the previous 99% confidence upper limit to $1.3 \times 10^{12} M_\odot$ for isotropic orbits (Zaritsky et al. 1989). Zaritsky et al. concluded that the discrepancy with the earlier estimates arose because the initial sample was biased toward systems with smaller mean separation. Importantly, the distribution of velocities among the satellites including Leo I, the most distant satellite which happens to account for the bulk of the difference between the mass estimates, is consistent with both the hypothesis that the satellites are distributed within a massive isothermal halo and with mass estimates using other Local Group galaxies (cf. Zaritsky et al. 1989). However, one object alters the mass estimates to such a large degree that one can argue that Leo I is unbound to the Galaxy and should not be included in the analysis. A strong counter argument is difficult to mount, and it may be impossible to significantly enlarge the sample. The major avenue for progress in this field lies in the measurement of transverse velocities, which for a majority of satellites should be completed within the next decade. However, even with known space velocities, it may be complicated to derive the satellite orbits.

Given positions, velocities, and an assumption of how satellites are distributed along their orbits one can convert the observations into estimates of the depth of the gravitational potential. Typically, one assumes that the particles are randomly distributed among their orbits. However, the assumption that the orbital phases of satellites are uniformly distributed, which is adopted by most investigators, is unjustified. Minimum mass timing argument calculations show that the farthest satellites are only in their second orbit. Timing arguments have a distinct advantage over other techniques because they incorporate other knowledge of the Universe, for example the age, and do not presume random orbital phases. Timing arguments produce estimates of the mass of our galaxy that are slightly larger than derived otherwise. It may be important to incorporate the advantage of timing arguments into analyses of other satellite samples.

3. COMPANIONS OF OTHER GALAXIES

3.1. Binary Galaxies

The initial studies of companion galaxies concentrated on binary galaxies because those are the simplest to identify and observe. In Figure 2 I have plotted, constrained within 500 kpc projected separation and 500 km s^{-1} velocity difference, the data from four of the most recent studies. As is evident from this figure, the data are not strikingly different among the various samples. The

White et al. sample has some binaries with large velocity differences and large separations simply because it is a bigger sample. In general, the data are concentrated within 100 kpc projected separation and within 200 km s^{-1}. Therefore, differences among conclusions from the various studies must be due to interpretation.

To illustrate the discordance among various studies let us consider the conclusions from some previous studies. Except for the results of Karachentsev's 1972 study and Turner's 1976 study, the estimated M/L's vary by at most a factor of 3 among the studies (Table 1: $H_0 = 75 \text{km s}^{-1} \text{Mpc}^{-1}$ used throughout). However, the conclusions from various studies are dramatically different. For example, van Moorsel (1987) concluded that a point mass model was inadequate even for companions at large separations (100 to 200 kpc). On the other hand, Schweizer (1987) concluded that point mass models provide a good approximation for the orbits of companions. Erickson et al. (1987) concluded that no additional mass to that inferred from the rotation curve was necessary to explain the dynamics, but White et al. (1983) concluded that there are extended halos consistent with a flat extrapolation of the rotation curve out to radii of about 100 kpc.

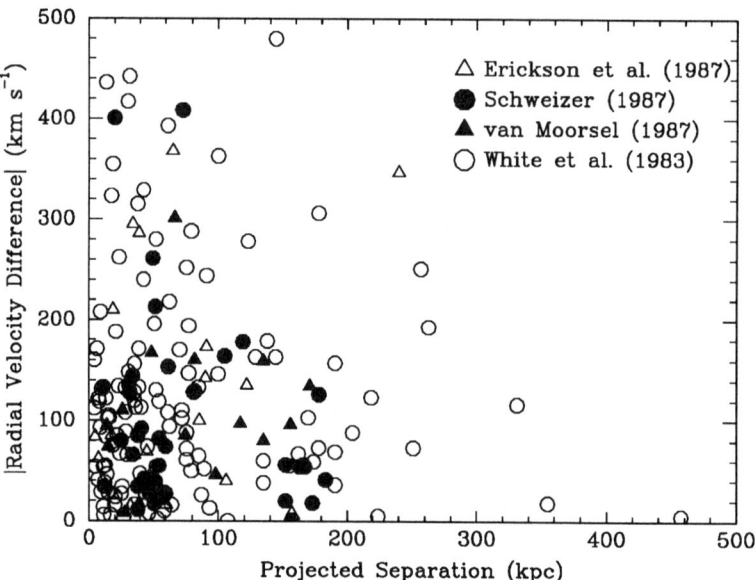

FIGURE 2. Comparison of Data from Previous Studies. The bulk of the data is clearly clustered in the lower left corner of the figure. A few data outside the bounds of the figure are not shown.

The root of the variance among the studies is that the quantitative results (i.e. M/L's) are not a conclusive differentiator among the various models because the data do not probe sufficiently large separations. Studies that concentrate on companions with projected separations smaller than 100 kpc and on large spiral primaries, which have flat rotation curves out to at least several tens of kpc, will

TABLE 1. Results from Binary Galaxy Studies[a]

Sample[b]	P52	K72	T76	P79	WHLD83[c]	S87	vM87
M/L	50	8	98	50	28	32	15

[a] These numbers are intended for rough comparison only. Model assumptions and sample selection are important, so one should refer to the original reference before adopting any particular M/L. Holmberg (1937) found similar masses as P52, but did not calculate M/L. EGH87 also do not present a global M/L value for their sample, and so their result has not been placed in the Table. All M/L values have been converted to $H_0 = 75 {\rm km\,s^{-1}\,Mpc^{-1}}$. There has been no attempt to select results from models of similar characteristics, so the comparison is for illustrative purposes only.
[b] Samples are referred to as follows: P52 (Page 1952), K72 (Karachentsev 1972), T76 (Turner 1976), P79 (Peterson 1979), WHLD83 (White et al. 1983), S87, Schweizer 1987, vM87 (van Moorsel 1987).
[c] Value presented is for r = 100 kpc, isotropically distributed velocities, and for L_* galaxies.

produce intermediate M/L estimates regardless of whether the halo is truncated near the end of the observed rotation curve or is isothermal (i.e. flat rotation curve to arbitrarily large radii). Consider an L^* galaxy ($L_B^* = 2.06 \times 10^{10} L_\odot$; Efstathiou et al. 1988) with a circular velocity of $\sim 170 {\rm km\,s^{-1}}$ and extrapolate the flat rotation curve out to 100 kpc. The enclosed mass at 100 kpc for this galaxy is only $6.4 \times 10^{11} M_\odot$ and is only $M/L_B = 31$. It is far more interesting to examine the potential at $r > 100$ kpc where the differences between different mass models become significant. In fact, in the analysis of the dynamics of satellites of our own galaxy we saw that a definitive case for a massive halo could only be made after including satellites out to distances beyond 200 kpc. One sample of binaries that does probe large radii is that presented by Charlton and Salpeter (1991), who concluded that galaxies are surrounded by large halos. Unavoidably, when companions at large separations are included in the analysis, contamination by galaxies that are physically unassociated with the primary, but appear close in projection, becomes a serious problem.

3.2. Contamination

As in the study of satellites of our own galaxy, it is not evident which companions are gravitationally bound to the primary and which are not. However, in external systems only projected separations are observed, and so the problem is exacerbated. These apparent companions, which are simply projected background galaxies, have no physical connection to the primary and only serve to dilute the information contained in any sample of companions galaxies. To illustrate the effect of interlopers I present the application of the Bahcall and Tremaine point mass estimator (Bahcall and Tremaine 1981) to the Schweizer sample of binary galaxies (Schweizer 1987). By applying the estimator to the entire sample (43 pairs) I derive an average mass per pair of $4.1 \times 10^{12} M_\odot$. If I remove the two pairs with $|\Delta v| > 500 {\rm km\,s^{-1}}$, under the presumption that these are not true pairs, then the estimated mass per pair decreases to $2.1 \times 10^{12} M_\odot$. Furthermore, if the two pairs with the next largest values of $r_p(\Delta v)^2$ are removed, then the estimated mass decreases further to $1.3 \times 10^{12} M_\odot$. Therefore, even if interlopers comprise less than 10% of the sample, the estimated mass per pair may be overestimated by a factor of four. This example illustrates that it is imperative to minimize contamination, to accurately estimate the contamination fraction, and to devise analysis techniques that account for interlopers. Most investigators address the problem by removing from the sample those objects which they believe are especially likely to be interlopers. However, a model that includes a representation of interloping galaxies is particularly desirable because it would eliminate the need to arbitrarily prune the sample.

3.3. Satellite Galaxies

As mentioned above, it is preferable to have test objects that are less massive than the primary. A recent study of companion galaxies focused on satellite galaxies (Zaritsky et al. 1993). The projected separations and velocity differences from that sample is plotted in Figure 3 and the luminosity function of the satellites is plotted in Figure 4. Note that large radii are much better sampled than in previous binary galaxy studies (cf. Figure 1) and that the majority of the companions are fainter than the LMC. Both of these aspects should allow a more robust determination of halo parameters. However, companions at such large separations are unlikely to be dynamically relaxed within the potential of the primary. As mentioned earlier, Leo I, the furthest satellite of the Galaxy, is estimated to be currently on only its second outward passage. Therefore, it becomes necessary to develop models that do not contain the implicit assumption that the test objects are dynamically relaxed.

By using simple models of the formation of galactic halos, the dynamics of simulated satellites drawn from a variety of potentials which form in universes of different Ω_0 can be compared to the data. Models valid for universes with critical density have been investigated by many including Gott (1975) and Gunn (1977). Complete analytic formulations that include the effect of relaxation were presented by Bertschinger (1985) and Fillmore and Goldreich (1984). To investigate halos in open universes White and Zaritsky (1992) developed analogous models. The scale imposed by the last shell of bound material necessitates computer generated solutions. In those computer simulations White and Zaritsky also included artificial generation of angular momentum, which then allowed them to investigate the dynamics for both a variety of orbital eccentricities and Ω_0's.

Zaritsky and White (1992) use these models and an adopted uniform spatial distribution of interlopers to derive confidence limits on Ω_0 and orbital eccentricities through a maximum likelihood analysis. By adopting a model for the distribution of interlopers and by using an estimate of the interloper fraction (from Zaritsky 1992), they can self-consistently address the contamination problem. They conclude that all models with $\Omega_0 < 0.03$ can be rejected with more than 99% confidence. Solid lower limits on the mass and extent of halos can be deduced by examining halos that form in a universe with $\Omega_0 = 0.03$. These halos, which can be rejected at with high confidence, have $1.1 \times 10^{12} M_\odot$ within 200 kpc ($M/L_B \sim 80$) and at least 20% more mass than that out to 500 kpc. Best fit

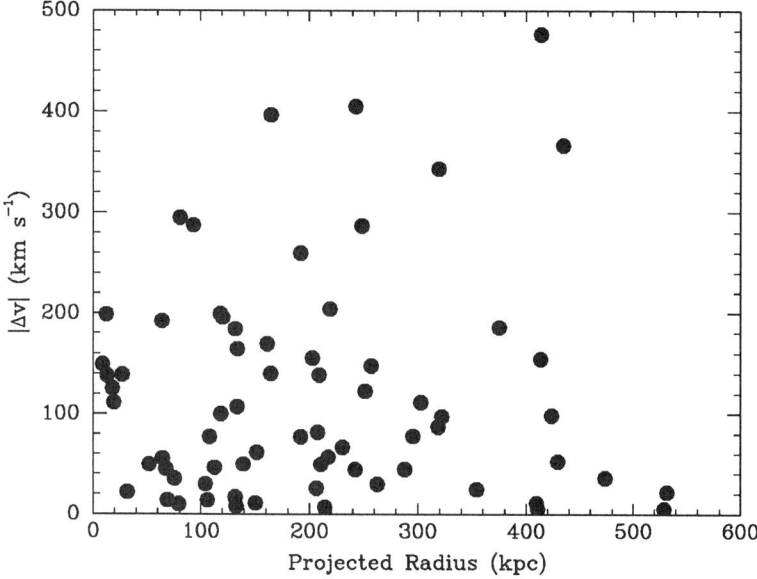

FIGURE 3. $|\Delta v| vs. r_p$ for Satellites from the Zaritsky et al. Sample.

models produce estimates of Ω_0 that are between 0.1 and 0.3, although models with $\Omega = 1$ are acceptable. It seems certain, provided we properly understand gravity at large separations, that the halos of spiral galaxies are dominated by dark matter and that these halos extend well beyond 200 kpc. We are currently enlarging the dataset to strengthen our statistical confidence of the results and to study subtler issues of the dynamics of satellites in galaxy halos.

3.4. What Does the Future Hold?

As always, more data is clearly the first step. Not only would more data tighten the statistical confidence limits, but also address other questions presented by Zaritsky et al. (1992). One particularly interesting but preliminary result they presented is that there appears to be slight net rotation of the satellites in the same direction as the disk. If additional data support this result, this observation would provide information for galaxy formation models and serve as input to future models of satellite dynamics.

One particular weakness of the existing models is that the generation of angular momentum is artificial and all external gravitational influences are ignored. Soon, cosmological simulations will have sufficient resolution to investigate both the formation of large and small scale structure. These simulations will provide important constraints on valid values of orbital eccentricities and will quantify the effects of nearby companions. The latter will enable future investigators to understand better the effects of certain isolation criteria.

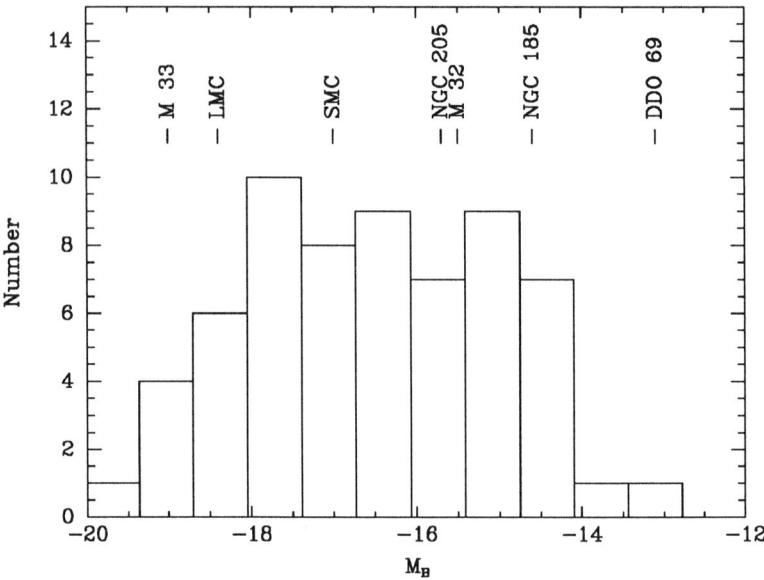

FIGURE 4. Luminosity Function of Satellites from the Zaritsky *et al.* Sample. Positions of Local Group members are labeled for comparison.

4. SUMMARY

Current studies seem to have resolved the question of whether spiral galaxy halos are truncated within a hundred kpc or whether they are extended beyond several hundred kpc. Dynamical estimates of both the mass of our galaxy and those of other galaxies indicate that spiral galaxies are surrounded by halos that extend beyond 200 kpc and contain at least $10^{12} M_\odot$. Note that the lower bound on Ω_0 from models of satellite dynamics (0.03) is marginally outside the range of acceptable $\Omega_{0,baryons}$ from standard big bang nucleosynthesis ($0.017 < \Omega_{0,baryons} < 0.028$; Olive *et al.* (1990)). Although they are very large, the halos are not isothermal spheres because of the finite age of the Universe. The mass profile is complex and must be investigated with realistic models of the formation of halos. This has now been done with simple models of halo formation. In models that correspond to the derived lower bound on Ω_0, the mass increases nearly linearly within 200 kpc and begins to increase more slowly beyond 200 kpc. Refinement of modeling and larger samples may make possible the measurement of the mass distribution to larger radii and as a function of the Hubble type of the primary.

Many of the tenets presented in this discussion are just as valid when discussing the dark matter distribution within groups of galaxies. The dynamics are more complex, but it is still true that the orbital phases will not be uniformly distributed (i.e. the systems are not relaxed) and that contamination will be important. In fact, contamination may be a more serious problem in study of groups, since groups are defined as an excess of galaxies in a region of the sky and are of larger angular extent than single galaxies. Furthermore, the estimator techniques, which are typically applied to groups, are extremely sensitive to interlopers. I suspect that there are still serious systematic difficulties with the estimated M/L ratios of groups and that the many-body dynamics involved in untangling these problems will not allow a proper treatment of this problem until cosmological simulations with high resolution that cover much of parameter space are available.

Acknowledgements: The author gratefully acknowledges invaluable input into the formulation of the ideas discussed in this paper from his principal collaborator on much of the work described here, Simon White. This work was supported by NASA through grant HF-1027.01-91A from STScI, which is operated by AURA, Inc., under NASA contract NAS5-26555.

5. REFERENCES

Bahcall, J.N., and Tremaine, S. 1981, *ApJ*, **244**, 805.
Bertschinger, E. 1985, *ApJS*, **58**, 39.
Casertano, S., and van Gorkom, J.H. 1991, *AJ*, **101**, 1231.
Charlton, J.C. and Salpeter, E.E. 1992, *ApJ*, **375**, 517.
Erickson, K., Gottesman, S.T., and Hunter, J.H., Jr. 1987, *Nature*, **325**, 779.
Fillmore, J.A, and Goldreich, P. 1984, *ApJ*, **281**, 1.
Gott, J.R. III 1975, *ApJ*, **201**, 296.
Gunn, J.E. 1977, *ApJ*, **218**, 592.
Hartwick, F.D.A., and Sargent, W.L.W. 1978, *ApJ*, **221**, 512.
Karachentsev, I.D. 1972, *Catalogue of Isolated Pairs of Galaxies in the Northern Hemisphere*, Soobsch. Special Astrophysics Obs., **7**, 3.
Lorrimer, S.J., Frenk, C.S., Smith, R.M., White, S.D.M., and Zaritsky, D. 1991, in preparation.
Olive, K.A., Schramm, D.N., and Steigman, G. 1990, *Phys.Rev.B*, **236**, 454.
Olszewski, E.W., Peterson, R.C., and Aaronson, M. 1986, *ApJ*, **302**, L45.
Page, T. 1952, *ApJ*, **116**, 63.
Peterson, S.D. 1979, *ApJ*, **232**, 20.
Salucci, P., and Frenk, C.S. 1989, *MNRAS*, **237**, 247.
Schweizer, L.Y. 1987, *ApJS*, **64**, 427.
Turner, E.L. 1986, *ApJ*, **208**, 304.
White, S.D.M. 1981, *MNRAS*, **195**, 1037.
White, S.D.M., Huchra, J., Latham, D., and Davis, M. 1983, *MNRAS*, **203**, 701.
White, S.D.M., and Zaritsky, D. 1992, *ApJ*, **394**, 1.
van Moorsel, G.A. 1987, *A&A*, **176**, 13.
Zaritsky, D. 1992, *ApJ*, **400**, 74.
Zaritsky, D., Olszewski, E.W., Schommer, R.A., Peterson, R., and Aaronson, M. 1989, *ApJ*, **345**, 759.
Zaritsky, D., Smith, R., Frenk, C.S., and White, S.D.M. 1992, *ApJ*, **405**, 464 (ZSFW).
Zaritsky, D., and White, S.D.M. 1992, in preparation.

Faint Dwarf Galaxies in the Sculptor and Centaurus A Groups

STÉPHANIE CÔTÉ AND KEN FREEMAN
Mount Stromlo and Siding Spring Observatories, The Australian National University, Private Bag, Weston Creek Post Office, ACT 2611, Australia

ABSTRACT: An HI survey in the regions of the Sculptor group and Centaurus A group has yielded 15 and 21 dwarf galaxy members of these groups respectively. These nearby groups of galaxies (at distances of about 2.5 and 3.3 Mpc) offer a unique opportunity to study the very faint end of the galaxy luminosity function.

1. OBSERVATIONS AND FIRST RESULTS

The Sculptor Group is composed mainly of late-type spiral galaxies; its five major members (NGC 55, NGC 247, NGC 253, NGC 300 and NGC 7793) are all normal gas-rich spirals. In the Centaurus A group, all of the brightest members are abnormal (ie. show some form of activity): NGC 5128 (Centaurus A itself) is a peculiar radio galaxy, and NGC 5236 (M83), NGC 5253, and NGC 5102 all appear to be in a post or present starburst phase. There have been many suggestions in the literature that these peculiarities are induced by accretion of gas-rich dwarf systems (e.g., van Gorkom et al. 1990).

Our search for dwarf galaxies in these groups was done by visually inspecting SERC J plates, covering an area of 35° x 35° for each group, and choosing faint, low-surface-brightness candidates. These candidates were then observed in HI with the 210ft telescope at Parkes to obtain their redshift. A total of 15 dwarf galaxies (of which 6 were already known) were thus confirmed to be members of the Sculptor group, with heliocentric velocities ranging from 53 to 406 km s^{-1}. For the Centaurus A group the total is 21 (with 13 previously known), with velocities between 122 and 747 km s^{-1}. Optical spectroscopy was also carried out on these objects with the Siding Spring 2.3m telescope to confirm the HI redshifts, because of possible confusion in the beam from high-velocity clouds associated with the Galaxy.

The dwarfs in each group show a wider spread in their spatial and velocity distributions than do the brighter members of the groups, as is the case for the dwarfs in the Virgo Cluster (Bothun et al. 1985). In the Centaurus A group, there are two clumpy subgroups of dwarfs surrounding the two most massive members, NGC 5128 (Centaurus A), and M83. The dwarf populations of the two groups are not dynamically well mixed; the crossing times are 0.6 H_o^{-1} for the Sculptor group and 0.5 H_o^{-1} for the Centaurus group. Both groups are not yet virialized.

In the Virgo cluster, while the bright galaxies have luminosity functions that are closely gaussian, the dwarf irregulars have a luminosity function fitted by a Schechter function with a maximum at $M_B = -16.1$ (Sandage et al. 1985). Preliminary results from our CCD imaging of the dwarfs in both groups, with the Siding Spring 2.3m telescope, show an ever-increasing luminosity function to $M_B = -11.5$ (even without any correction for incompleteness). These results indicate that small groups of galaxies have large populations of faint gas-rich dwarfs.

Acknowledgement: This work was done in collaboration with Claude Carignan and Peter Quinn.

2. REFERENCES

Bothun, G., Mould, J., Wirth, A., and Caldwell, N. 1985, *AJ*, **90**, 697.
Sandage, A., Binggeli, B., and Tammann, G.A. 1985, *AJ*, **90**, 1759.
Van Gorkom, J.H., van der Hulst,J.M., Haschick,A.D., and Tubbs, A.D. 1990, *AJ*, **99**, 1781.

HI Mapping of Compact Groups

B.A. WILLIAMS
Princeton University,
Dept. of Astrophys. Sci. and Princeton University Observatory,
Peyton Hall, Princeton, NJ 08544-1001[1]

J.H. VAN GORKOM
Columbia University, Astronomy Department,
538 W. 120th St., New York, NY 10027

ABSTRACT: We present two new images of the neutral hydrogen (HI) in the direction of the compact groups of galaxies HCG (Hickson Compact Group) 23 and HCG 26. In HCG 23 the emission is associated with 3 of the spiral galaxies identified by Hickson as group members and with 5 other fainter galaxies that we have subsequently identified as new members. Emission is detected at the same velocity as that of HCG 23 over a region 4 times larger than that of the compact group. The H I morphology of HCG 23 would suggest that this compact group is part of a larger loose group of galaxies. In contrast, most of the emission from HCG 26 is contained within a single cloud feature as large as the entire compact group and nearly centered at the position of group's center. The change in the cloud's kinematics along its major axis suggest two separate dynamical systems that merge in velocity and position space.

1. INTRODUCTION

We have undertaken a long-range observational program to image in the 21-cm line all of the Hickson (1982) compact groups that have been detected in neutral hydrogen. The motivation for imaging these groups is that we hope the position and kinematics of the neutral hydrogen may give clues about the dynamical history of compact groups and some indication of their future evolution. To date, eleven groups have been imaged in the 21-cm line and we have been responsible for imaging 10 (HCG's 2, 16, 18, 23, 26, 31, 33, 44, 79, and 96) of these at the VLA while HCG 92 has been imaged at Westerbork (Shostak *et al.* 1984).

Data reduction is nearly complete for 6 of the groups. Based on their H I morphology, the few groups for which the data reduction is nearly complete fall into 3 or 4 categories. The first category consist of groups that appear to form compact configurations that have sizes of the same order of magnitude as normal galaxies. Within these groups, the individual members appear to be miniature stellar systems. HCG's 18 (Williams and van Gorkom 1988), 31, and 79 (Williams, McMahon, and van Gorkom 1991) are examples of the groups in this first category. In the second category (HCG's 44 [Williams, McMahon, and van Gorkom 1991], 92, and 26) are groups that have sizes that are 3–5 times larger than normal galaxies and the individual members within can be described as having normal dimensions. The third category includes compact groups that are parts of larger loose groups of galaxies. Our best example would be HCG

---

[1] on leave from the Department of Physics and Astronomy, University of Delaware.

23. A fourth category could easily be added if HCG 18 is not a compact group but a single irregular-type galaxy. This category would include those compact groups that have been misidentified and are in fact single galaxies.

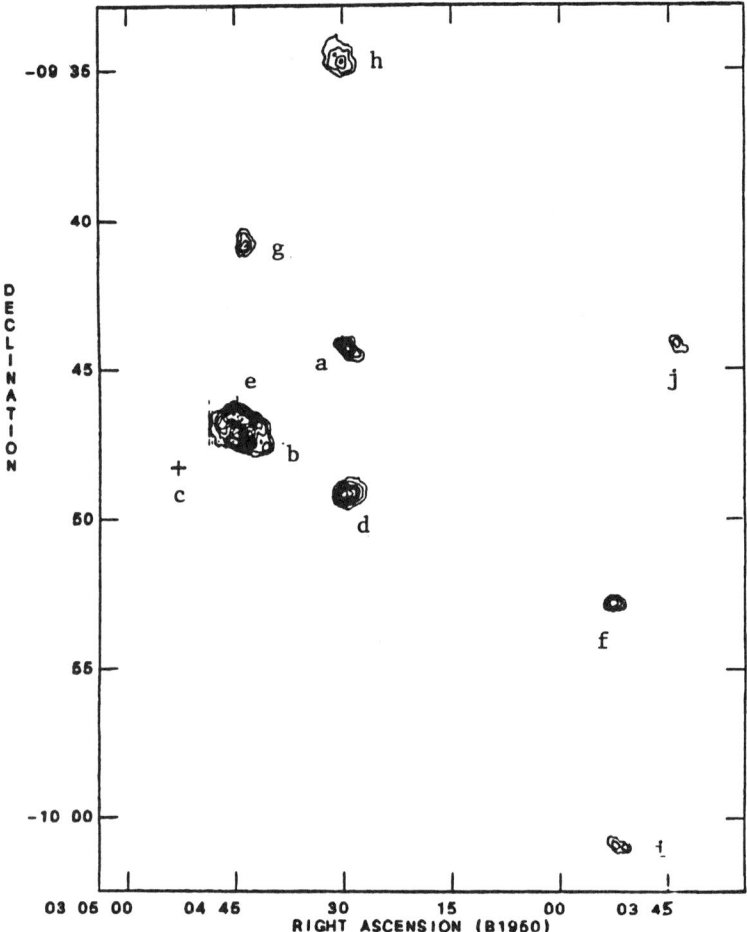

Figure 1. VLA integrated map of HI emission in HCG 23. The lowest contour level and contour interval are 0.05 Jy km s^{-1}. The peak emission is 0.48 Jy km s^{-1}. The beam size is 17"x21" at a position angle of 24°.

The HI morphology of the Hickson groups does confirm that the galaxies within these systems are physically associated. The total hydrogen ranges between 10^9 and 10^{10} solar masses. When the gas cannot be clearly associated with individual galaxies its extent is 1-3 times larger than the entire compact group. There are examples in which the gas is centered near the centroid of the group as well as large amounts which are displaced from the centroid formed by the members. In every case where the gas is detected, it is at the same radial velocities of the galaxies within the group and shows systematic motions over small scales comparable to the sizes of the miniature galaxies and over larger scales which encompass the entire compact group and beyond.

We report here our most recent results which include new images of the neutral hydrogen in HCG 23 and 26 along with a discussion of the distribution and kinematics of their gas properties.

2. DISTRIBUTION AND KINEMATICS OF THE GAS

The HI emission detected in the direction of HCG 23 is shown in Figure 1. As is the case with HCG 44, most of the integrated H I emission can be clearly associated with individual galaxies that lie in the primary beam. We detect the three brightest spirals out of the five members (**a** through **e**) identified by Hickson *et al.* (1989). Figure 1 shows that we have also detected five additional members that lie well outside of the compact group. Galaxy **j** is a bright Sab/Sb type galaxy while the other four detections (**g**, **h**, **f**, and **i**) appear as faint images that are barely discernible as galaxies on the Palomar Observatory Sky Survey. The spiral galaxies within the compact configuration show an asymmetry in their distribution and kinematics (Figure 2). In galaxy **a** most of the gas is located on one side of the optical position assuming its location is correct. Galaxy **b** seems to cut off abruptly at the position of galaxy **e**. Galaxy **d** is detected very close to 1400 MHz where interference occurs at the VLA. As a result 2 channels of data had to be removed at frequencies near galaxy **d** so some H I emission and velocity information are lost in these two integrated maps (Figure 2). All three velocity maps show an asymmetry in the motions of the gas relative to the optical centers. Rubin et al. (1991) have imaged HCG 23 in the H–alpha line and measured optical velocities along the major axis of the three brightest spirals. They also measure asymmetries in the motions of galaxy **a** relative to the nucleus. The agreement between the motions of the ionized gas (Rubin *et al.* 1991) and the neutral gas is remarkable in the case of galaxies **a** and **b**. HCG 23 appears to be part of a much larger system that extends over a region ≈ 500 kpc in which we have detected $\approx 10^{10}$ solar masses of neutral hydrogen in eight galaxies that have an average velocity of $\approx 4800 \mathrm{km\,s^{-1}}$ and a radial velocity dispersion of $180 \mathrm{km\,s^{-1}}$.

The H I emission in the direction of HCG 26 (Figure 3) is contained within two cloud features. The strongest emission is centered on the four northern most members, **a**, **b**, **d**, and **g**. A CCD images of the group reveals that galaxy **b** is not a separate system but simply luminous material that extends outward from galaxy **a** (private communication, Steve Zepf 1992). The weaker cloud feature to the south appears to be associated with galaxy **e** which is viewed nearly edge–on and is typed as an irregular galaxy (Hickson *et al.* 1989). Figure 4 shows the large–scale systematic motions that occur across the entire cloud along the same direction as the position angle of the disk associated with the brightest edge-on spiral in the group. The optical velocities measured by Hickson at the positions of galaxies **a** and **b** are again very similar to velocities of the gas at these same

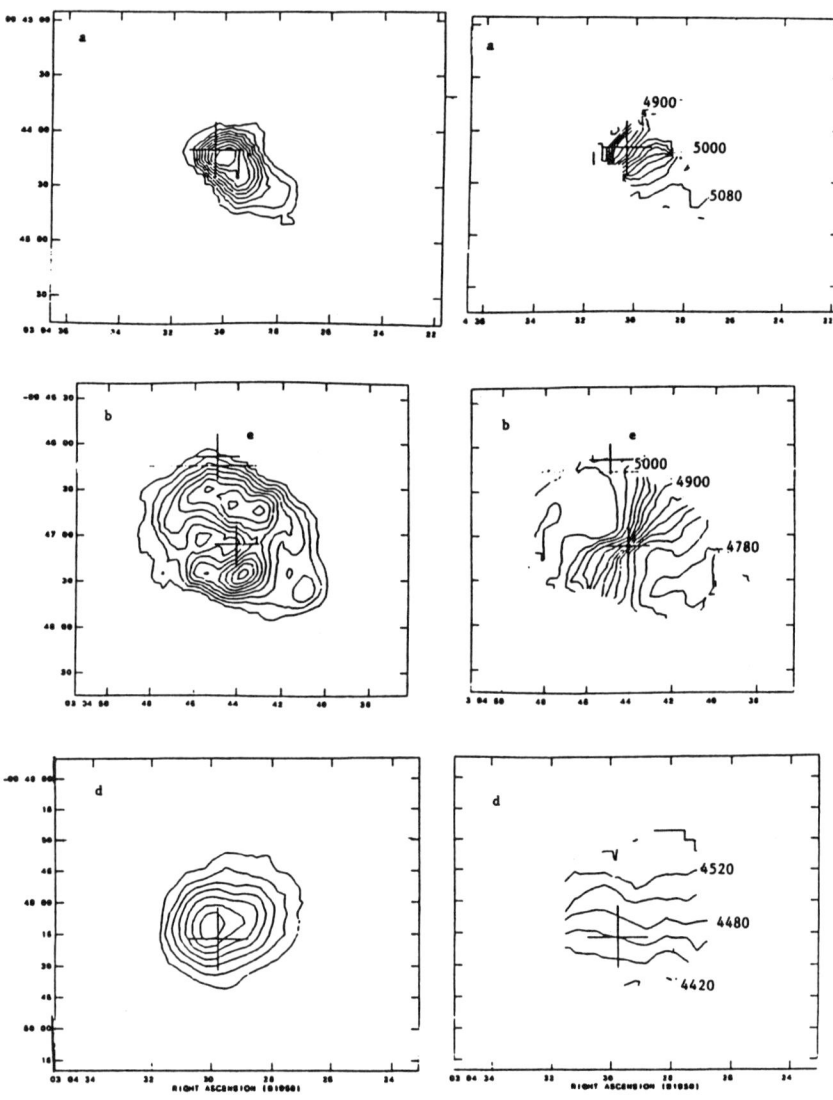

Figure 2. Detailed integrated map of HI emission in galaxies **a**, **b**, and **d** from Figure 1 along with the corresponding velocity fields (km s^{-1}) in intervals of 20 km s^{-1}.

positions. Systematic motions are also observed in the smaller cloud to the south. The two clouds appear to merge almost continuously in velocity and position space. The velocity dispersion of the larger cloud is ≈ 370km s^{-1} while that of the smaller cloud is only ≈ 80km s^{-1} which seems typical for irregular type galaxies (Gallagher and Hunter 1984). Both cloud features are much larger than the individual galaxies with which they are associated.

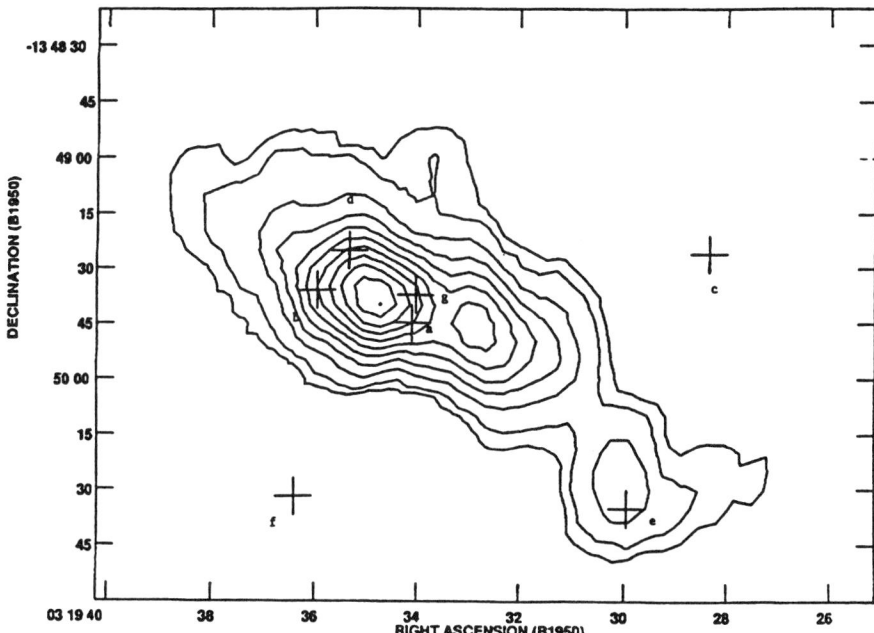

Figure 3. VLA integrated map of HI emission in HCG 26. The lowest contour is 0.05 Jy km s^{-1}. The next contour level and contour intervals thereafter are 0.10 Jy km s^{-1}. The peak emission is at 1.0 Jy km s^{-1}. The beam size is 25"x17.6" at a position angle of 6°.

3. SUMMARY

The number of compact groups that have been imaged is still very small to draw any real conclusions about the sample of Hickson groups. We will continue to map all compact groups that have been detected in neutral hydrogen hoping to catch groups in different stages of the merging process if this is indeed the fate of most compact groups. The different stages of merging might be described by their gas morphology, i.e. its distribution and kinematics. As we continue to add new groups to those that have been mapped a clear picture may emerge as to how compact groups evolve into single galaxies and what happens to their neutral gas.

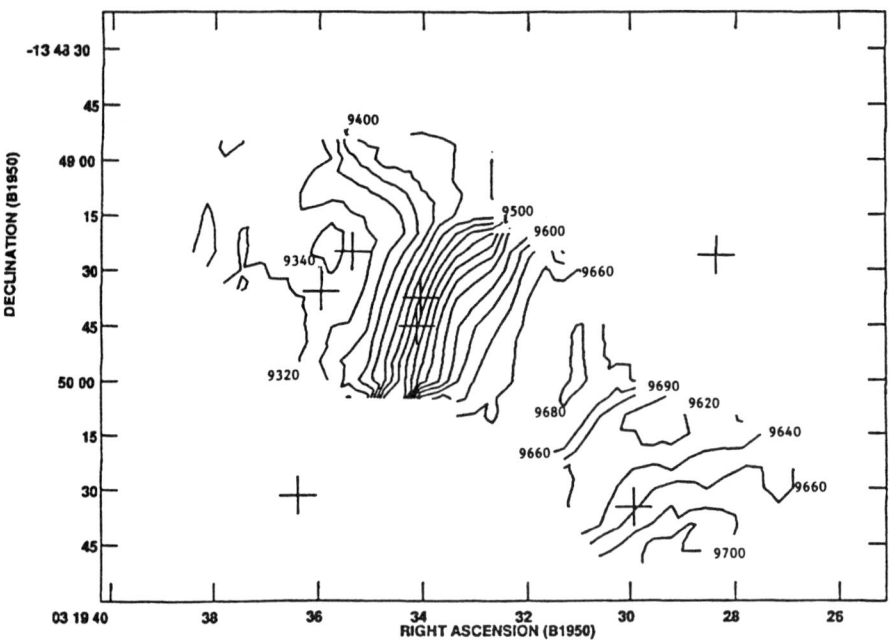

Figure 4. Map of the HI velocity field (km s^{-1}) for the HCG 26 cloud in intervals of 20 km s^{-1}.

4. REFERENCES

Gallagher, J.S. and Hunter, D.A. 1984, *ARA&A*, **22**, 37.
Hickson, P. 1982, *ApJ*, **225**, 382.
Hickson, P., Kindl, E., and Auman, J. 1989, *ApJS*, **70**, 687.
Rubin, V., Hunter, D., and Ford, W.K. 1991, *ApJS*, **76**, 153.
Shostak, G.S., Sullivan, W.T., and Allen, R.J. 1984, *A&A*, **139**, 15.
Williams, B.A., and van Gorkom, J.H. 1988, *AJ*, **95**, 352.
Williams, B.A., McMahon, P.M., and van Gorkom, J.H. 1991, *AJ*, **101**, 1957.
Zepf, S. 1992, private communication.

Compact Groups: Observations and Theories

GARY A. MAMON

DAEC, Observatoire de Meudon, 92195 Meudon, FRANCE

ABSTRACT: The present status of compact groups is reviewed, from both observational and theoretical viewpoints. A wealth of recent observational studies point to a high level of dynamical interactions between galaxies in compact groups. However these interactions are not as frequent as observed in catalogs of isolated pairs of galaxies or inferred from theoretical studies. This moderate level of interactions is difficult to understand if compact groups are bound dense systems of galaxies. New N-body simulations of loose groups of galaxies confirm that 1D chance alignments of galaxies are at least ten times more probable than 3D bound dense subgroups, and are rich in binary galaxies. One then expects the level of observed interactions to be situated between that of field galaxies and that of interacting binary galaxies. Alternative theories of compact groups are critically examined.

1. INTRODUCTION

Compact groups of galaxies correspond to the isolated systems of highest apparent density in the Universe, with densities comparable to or greater than those of the cores of rich clusters of galaxies. As such, these systems should be the ideal place to look for galaxy interactions. Indeed, an increasing number of observational studies have searched for signs of dynamical interaction in compact groups usually with some success. Yet, despite the vast amount of data compiled on these systems, there is still no agreement on whether compact groups are as dense as they appear to be. I review here the status of the observational studies of compact groups, and continue with an analysis of the different theories competing to explain these systems.

2. CATALOGS

Three catalogs of compact groups are frequently discussed in the literature: the *Shakhbazyan*, *Rose*, and *Hickson* catalogs, all of which are selected from visual inspection of the POSS plates. We discuss each in turn.

Shakhbazyan (1973) published a list of 30 *compact groups of compact galaxies*, with the following constraints:

- $5 \leq N <$ "rich clusters",

- "sufficiently isolated and compact group", and

- most galaxies must be "compact".

Subsequent studies (Baier and Tiersch 1979, and references therein) have extended the list to a total of 377 compact groups over 17% of the sky. Note that the selection criteria for Shakhbazyan compact groups (hereafter, SCGs) are very vague. There is no quantitative definition of compactness or isolation.

Moreover, the SCGs have the peculiarity of having mostly compact, high surface brightness galaxies, which is not the case in the other catalogs presented below.

The first compact group catalog to incorporate a quantitative definition of compactness is due to Rose (1977). His criteria are

- $N \geq 3$,
- $B \leq 17.5$, and
- $\Delta\Omega \, \mathcal{N}(B_{\text{faint}}) \leq 0.0035$.

where $\Delta\Omega$ is the solid angle of the smallest polygon delimiting the galaxy centers, and $\mathcal{N}(B_{\text{faint}})$ is the projected number density of galaxies in the field up to a faint magnitude limit B_{faint}. Restating the last (compactness) criterion, the Rose groups (hereafter, RCGs) represent projected number density enhancements of $860N/3$ over the field, where N is the group membership. Altogether, there are 205 RCGs in 7.5% of the sky, among which 170 are triplets. The RCG catalog suffers from having few quartets and higher multiplicity systems and lacks an isolation criterion.

The Hickson (1982) catalog of compact groups (hereafter, HCGs) is the first one to implement quantitative criteria for both compactness and isolation, according to:

- $N \geq 4$,
- $m - m_1 \leq 3$,
- $\mu_E \leq 26.0$, and
- $N_{\text{annulus}} = 0$.

Here the compactness criterion refers to the red (E) POSS plates and is computed within the smallest circle containing all of the galaxy centers, while the isolation criterion refers to a concentric annulus whose radius is three times that of this circumscribed circle, and to the same faint magnitude limit as for the group. Hickson's catalog contains exactly 100 compact groups in 67% of the sky.

3. OBSERVATIONS OF HICKSON'S COMPACT GROUPS

Because Hickson's catalog is the best defined, I will now focus exclusively on these systems.

3.1. Stellar content

Optical studies of HCGs have pointed towards numerous features of dynamical interaction between the member galaxies. High quality CCD images of all HCGs have been obtained at CFHT and preliminary analysis indicates that one-third of the galaxies in HCGs show morphological disturbances (Hickson 1990), while in one-third of the groups there are three or more galaxies in interaction (Hickson 1990, in the subsequent discussion). Assuming that virtually no HCGs have only *one* galaxy showing signs of interaction, then these two fractions imply that at least 40 of the 100 HCGs have *zero* interacting galaxies. Tikhonov (1990) finds a similar fraction of interacting galaxies from his inspection of plates obtained at the 6m telescope, but only 12% of the 59 HCGs that he observed contain 3 or more interacting galaxies, and roughly 40% have zero interacting galaxies.

Surface photometry of ellipticals in HCGs yield normal $r^{1/4}$ law surface brightness profiles (Bettoni 1992; Perea 1992), and detailed surface photometry by Zepf (1992) shows that although HCG ellipticals show more often irregular than regular isophotes, their is a lack of ellipticals with boxy isophotes, compared to field and cluster ellipticals (Bender *et al.* 1988). If boxy isophotes are signatures of recent merging (Binney and Petrou 1985), then HCGs show less merging than other environments.

Roughly 7% of HCG spirals show tidal tails, which are another signature of strong dynamical interaction, whereas in the field Toomre (1977) estimates only 0.3% of spirals to have such features.

Another sign of merging could be the existence of elliptical galaxies with blue colors that trace a recent burst in star formation. Zepf, Whitmore and Levison (1991) find that 5.5% of the ellipticals in HCGs have $B - V$ colors that are 0.1 mag bluer than standard ellipticals of the same luminosity using data from Burstein *et al.* 1987. In contrast in that data set, only 2.5% of the ellipticals displayed blue colors in the same sense. Zepf and Whitmore (1991) have attempted to estimate the frequency of occurrence of such blue ellipticals with a simple Monte-Carlo merger model, but obtain 15% assuming that mergers occur at the rate of 1 per Gyr, roughly as expected if HCGs are bound dense systems. Moreover, the blue ellipticals found by Zepf *et al.* (1991), as well as Bettoni (1992) and Moles (1992), are low luminosity ellipticals, whereas one expects merger remnants to be giant ellipticals.

Moles (1992) finds that HCG ellipticals have $U - B$ colors that are 0.1 bluer than for standard galaxies (for given $B-V$), again suggesting that these galaxies are the sites of star formation. However, this effect is much weaker than what he sees in isolated pairs.

The optical luminosity function of HCG galaxies is a topic of some debate. One can either try to model the selection effects on the luminosities of the galaxies that build the groups (Mendes de Oliveira and Hickson 1991) or alternatively, consider the luminosities of the galaxies in the HCGs without regard to these selection effects (Sulentic 1992). In both studies, the luminosity function is fit to a Schechter function: $\phi(L)dL \sim (L/L_*)^\alpha \exp(-L/L_*)dL$. Mendes de Oliveira and Hickson find a shallow faint-end slope of $\alpha = -0.3$, consistent with a preliminary study by Heiligman and Turner (1980), while Sulentic finds a faint-end slope of $\alpha = -1$, consistent with a similar study by Hickson *et al.* (1984), and similar to that of field galaxies (Efstathiou, Ellis and Peterson 1988; Loveday *et al.* 1992 using the APM survey). Thus, there is disagreement on whether or not HCGs lack dwarf galaxies. However, one would expect that, by removing the important selection effect of limiting the magnitude range to 3 mags, Mendes de Oliveira and Hickson should have obtained a steeper slope than Sulentic.

3.2. Neutral gas

Williams and Rood (1987) performed the first systematic study of HCGs in 21cm, and found a 50% deficiency of HI gas in HCGs relative to galaxies in loose groups. Unfortunately, the reference loose group sample used by Williams and Rood was small (18 groups) and never published.

A finer study involving mapping the distribution of HI in HCGs, still in progress, has revealed so far show that the distribution of HI gas can be either associated with the individual spiral galaxies in the group, or may encompass the whole group (Williams *et al.* 1991), who suggest that the HI distribution may thus be a tracer of the age of the group, as the neutral gas will merger into a single cloud in more evolved groups (see Williams 1992).

3.3. Hot gas

X-ray data obtained with the EINSTEIN observatory (Bahcall, Harris and Rood 1984) shows hot gas associated with individual galaxies (HCG 16), but also distributed throughout the group (HCG 92) and perhaps centered on a region of radio continuum emission (see §3.4). Note that diffuse X-ray emission was also seen in a relatively dense (5 galaxies within just twice the size of a typical HCG) loose group (Biermann, Kronberg and Madore 1982). Analysis of ROSAT data is underway (van Gorkom and Williams 1992). Unfortunately, contrary to HI line studies, X-ray studies are purely 2D, and thus subject to projection effects.

3.4. Warm gas, dust, star formation, and nuclear activity

Menon and Hickson (1985) and Menon (1990) found that when 6 and 18 cm radio sources in HCGs are associated with ellipticals, these are almost always the brightest group member, whereas no such trend is seen when such radio sources are associated with spiral galaxies.

Otherwise, Allen and Hartsuiker (1972) discovered a ridge of continuum radio emission between two galaxies in Stephan's Quintet (HCG 92), which may correspond (van der Hulst and Rots 1981) to the region at the interface of the interstellar medium of one galaxy with the intergalactic medium of the group, giving a burst of star formation (indeed H II regions lie in the ridge).

The far infrared properties of HCGs are also a matter of debate: Hickson *et al.* (1990) find that HCGs have far IR luminosities roughly a factor of 2 larger than expected from their optical luminosities, when compared to isolated spiral galaxies, a result contradicted by Sulentic and Mello Rabaça (1992). Such an IR excess is usually thought to signify an enhanced rate of star formation, although it is now believed that IRAS emission measures more than simply star formation (Thuan 1992).

Therefore, it is most interesting to notice that almost all HCG ellipticals and lenticulars show [N II] emission in their nuclei (Rubin, Hunter and Ford 1991), although this may be due to the high spectral resolution reached in this study. This nuclear emission is probably related to the ionized gas thought to produce the nuclear 6 and 18 cm emission seen by Menon and Hickson (1985) and Menon (1990).

3.5. Kinematics

The first systematic kinematical studies of HCGs are due to Williams et al. (1991) in 21cm, and Rubin et al. (1991) in Hα. Williams et al. show that the neutral gas is in ordered rotation in the spirals of two HCGs, but in one (HCG 31) the kinematics is quite complicated, probably related to the interaction of its member galaxies. Note however that HCG 31 is not a "real" HCG as its faintest member 5.09 mags fainter in the B band and 4.30 mags fainter in R from the magnitudes listed in Hickson, Kindl and Auman (1988). Moreover, the group is not isolated as it has a bright neighbor lying $1.03'$ from its center, while its angular radius is $0.445'$ from spherical trigonometry on the coordinates given in Hickson, Kindl and Auman. However, inclusion of this neighbor may make a new isolated compact group, although this remains to be checked.

Removing HCG 31 from the sample of HCGs studied by Rubin et al. (1990), 14 spirals out of 26 have rotation curves with small asymmetries and sometimes very strange features, such as sinusoidal patterns, suggesting that these galaxies are dynamically disturbed by close companions. This high level (54%) of interacting galaxies, compared to the 33% of galaxies whose images show signs of interactions (§3.1) suggests that kinematical analyses are more sensitive tracers of dynamical interactions than image analysis.

Finally, a kinematical study of 47 ellipticals in 10 RCGs and 25 HCGs reveals little evidence of kinematical distortions but the steep relation $L \sim \sigma_v^{6.2}$ (Bettoni 1992) instead of the usual Faber-Jackson (1976) $L \sim \sigma_v^4$. Moreover, Zepf (1992) finds that the giant ellipticals in HCGs tend to have velocity dispersions that are 20% lower than cluster galaxies of same size. This is well illustrated in Fig. 1 of Djorgovski, Weir and de Carvalho (1992), who show that the brightest HCG ellipticals have normal velocity dispersions but the other ones (though still fairly bright) have velocity dispersions up to 5 times lower than normal ellipticals of the same luminosity.

3.6. Environment

Various studies have argued that the environments of HCGs are usually too poor to understand HCGs being caused by chance alignments within loose groups or clusters (Williams and Rood 1987; Sulentic 1987; Rood and Williams 1989; Palumbo 1993). However, of the now 9 HCGs that have at least two galaxies included in galaxy catalogs from which group catalogs were assembled, 8 of these are embedded in loose groups, while only one *is* the (isolated) group (Mamon 1993). Moreover, the other studies disagree on which HCG is isolated, and among the 9 nearby ones mentioned above, many are called isolated by the four studies above. The difficulty comes from the fact that loose groups are not much denser than the field: indeed, Turner and Gott's (1977) loose groups, which are among the densest, are only $> 10^{2/3} = 4.6$ surface density enhancements over the field.

4. THEORIES FOR COMPACT GROUPS

On the basis of the numerous signs of dynamical interactions listed in §3, virtually the entire community of compact group observers reaches the conclusion that Hickson's compact groups are bound dense systems of four or more galaxies (*e.g.*, Williams and Rood 1987; Sulentic 1987; Hickson and Rood 1988; Moles 1992; Palumbo 1993). If HCGs are bound dense systems, they could be the remnants of primordial dense groups, or they may have formed recently. Alternatively, it has been argued that HCGs may not be physical systems of at least 4 galaxies, but would be caused by chance alignments of galaxies along the line-of-sight within loose groups (Mamon 1986) and clusters (Walke and Mamon 1989). While there has been considerable debate between the proponents of these two hypotheses, it would be fair to state all the other possibilities for explaining HCGs. Indeed, HCGs could represent *unbound* subsystems within loose groups, which only survive for roughly their crossing time (Rose 1979). Also, HCGs could simply be loose groups at the epoch of cosmological collapse. Moreover, there must be some HCGs that are caused by chance alignments *in the field*. Furthermore, some HCGs may simply be H II regions within a single galaxy (Williams and van Gorkom 1988; Tikhonov 1990). Finally, there is some indication that the most distant HCGs may be the bright-ends of rich clusters (Mamon 1993). Below, various arguments are presented for or against these different hypotheses.

4.1. Real groups or chance alignments: old arguments

In spite of the strong interactions seen in HCGs, there are many arguments working against the bound dense group hypothesis. First and foremost, it is difficult to see how such bound dense groups will form in the numbers observed today, which Mamon (1993) estimates at roughly 1% of the number of loose groups in a comparable portion of the Universe.

Statistical N-body simulations of dense groups show that of nearly 1000 simulated dense groups, none survive for a Hubble time (Mamon 1987), which thus seems to argue against the idea that HCGs are the lucky survivors of primordial dense groups. This is addressed further in §4.2.

Alternatively, a bound dense group would have to form by two-body processes, by which sufficient exchange of energy between the galaxies in a loose group may cause a bound dense subgroup to form inside. Unfortunately, no signs of such subsystems were seen in Mamon's (1987) statistical sets of dynamically simulated loose groups, although the database of simulated groups was still not large enough to conclude. This issue is addressed in §4.2 below.

If HCGs are bound and dense, there should be enough mergers to show a statistically significant increase in the mean difference of first to second rank galaxy magnitudes (Tremaine and Richstone 1977). This is indeed seen to occur in simulated dense groups, half-way between their virialized initial conditions and the time when they lose their HCG identity (Mamon 1987). In contrast, the HCGs show a normal $\langle m_1 - m_2 \rangle$ (Mamon 1986). See §4.2 below for more on this issue.

Moreover, two observational tests favor the chance alignment idea over the bound dense group one. First, HCGs do not obey the group/cluster morphology-density relation (Postman and Geller 1984), but follow instead a parallel relation, which, at a given density, makes them too spiral-rich (Mamon 1986). In contrast, because mergers should be operating in dense groups, and that these have the ideal low velocity dispersions in comparison with rich clusters, one would expect the HCGs to have a larger elliptical fraction than rich clusters (Mamon 1992). Alternatively, if their spiral fraction is indicative of their density, then HCGs would be 200 times less dense in 3D then what they appear in 2D (Mamon 1986). The chance alignment idea predicts a similar ratio and also predicts that the HCG relation is roughly parallel to the group/cluster one, because if HCGs are chance alignments, their morphological mix should mimic that of their parent system, while scaling arguments show that the denser parent systems will lead to more compact chance alignments (Mamon 1986).

Second, if HCGs are chance alignments within loose groups, then the ratio of HCG to loose group M/Ls should be equal to

$$\frac{R_{\rm HCG}/R_{\rm LG}}{N_{\rm HCG}/N_{\rm LG}} \simeq \left(\frac{200^{-1/3}}{4/8}\right) \simeq 1/3 \,, \tag{1}$$

where R and N refer to projected radius and number of galaxies, respectively, and where the system velocity dispersions and mean galaxy luminosities are assumed to be the same between chance alignment and parent loose group. Now Hickson (1990) gives the mean M/L of HCGs as roughly $40\,h$, whereas for loose groups one gets $125\,h$ (Tully 1987) and $190\,h$ (Ramella, Geller and Huchra 1989), and the ratio of these two M/Ls is in fine agreement with equation (1).

Finally, there has been some debate on the expected frequency of chance alignments within loose groups. Whereas dynamical simulations yielded values of a few percent (Mamon 1987), the static Monte-Carlo simulations of Hickson and Rood (1988) produced only $\sim$ a few 10^{-5} for typical loose groups. In an analytical study, Walke and Mamon (1989) explained this discrepancy as caused by the strong dependence of the frequency of chance alignments to the size of the parent system ($\sim R^{-4.5}$) coupled with the relatively small loose group sizes used in the dynamical simulations. As a consequence, the *median* frequency used in the Monte-Carlo study is an order of magnitude smaller than the more relevant *mean*. Moreover, the underlying model in the Monte-Carlo and analytical studies is a Poisson distribution of galaxies in the projected loose group, whereas in reality, the gravitational attraction of galaxies causes loose groups to be relatively rich in binaries. Now in their Monte-Carlo study Hickson and Rood show that adding one binary per group increases the frequency of chance alignments by 1.5 orders of magnitude, whereas, Tully's (1987) loose groups tend to have on average 1.5 binaries, defined in the same way (Walke and Mamon 1989) and one thus reaches (or surpasses!) the frequencies of 0.5% (see above) necessary to explain half of the HCGs as chance alignments (see Mamon 1993 for details).

4.2. Real groups or chance alignments: new arguments

First, HCGs may indeed be bound and dense, but forming at early epochs from the densest primordial density fluctuations. Governato, Bhatia and Chincarini (1991) reported simulating a dense quartet that kept their HCG appearance for a fifth of a Hubble time (after two galaxies merged and the group turned into a triplet). Now the results from this study do *not* indicate that the lifetime of HCGs is of the order of a Hubble time as was mentioned by Mendes de Oliveira and Hickson (1991), but one can extrapolate the results and wonder if HCGs may be the lucky survivors of primordial dense groups. The long merger time turns out, as pointed out by Governato et al., to be caused by their two most massive galaxies starting on a nearly circular orbit, thus maximizing the survival time against the merging of the pair. And cosmologically expanding initial conditions will not often give rise to the circular orbits required for enhanced dense group stability.

Because groups, as dense as HCGs appear to be, will condense out of their initial Hubble expansion in a short time, there is perhaps 10 times more time to form loose groups than to form primordial dense groups. One would have to invoke that primordial dense groups formed at a rate 10 to 100 times higher than loose groups today, which seems unlikely.

One can repeat the Tremaine-Richstone statistics mentioned in §4.1, but compute now the time average of the value of $\langle m_1 - m_2 \rangle$ between the initial conditions and the time they lose their HCG identity. This turns out to be statistically significantly large. So, if HCGs are primordial dense groups, they should have larger $\langle m_1 - m_2 \rangle$ (taking into account the HCG magnitude selection criterion $m - m_1 \leq 3$) than is observed, which thus seems to rule out that HCGs are primordial bound dense groups.

A series of new simulations of loose groups, based upon a the code in Mamon (1987) have been run to establish accurately the frequency of formation of bound dense subgroups within loose groups, and the frequency of binaries within chance alignments appearing in loose groups. The statistics on compact configurations is improved in three ways: 1) The simulations are run for 2.5 Hubble times instead of one. 2) The simulations are viewed (in projection along three orthogonal axes) 80 times per Hubble time, instead of 8. 3) For each set of parameters, 100 simulations are run (with different initial positions and velocities), instead of 20 to 50. Thus altogether, the statistics are typically 100 times better than previously.

The standard runs involve loose groups of initially 8 galaxies, sampled from a Schechter luminosity function of index -1 with a faint cut-off at $0.01\,L_*$ with the mean galaxy luminosity forced at $4.2\,h^{-2} \times 10^9\,L_\odot$ to within 1%, starting with random positions inside a sphere of radius $650\,h^{-1}$ kpc and random velocities satisfying on average the virial theorem. The dark matter is placed either mainly within galaxy halos, or in a common envelope. In both cases the group mass-to-light ratio is $120\,h$ to conform with Tully's (1987) NBG groups, while the galaxies have mass-to-light ratios of $20\,h$ without halos and $92\,h$ with halos.

The results of the new simulations are shown in Tables 1 and 2. The first important point from Table 1 is that 1D chance alignments are typically 10 times more frequent than 3D subgroups. If the dark matter is in individual halos, the dense subgroups are mostly bound, but if the dark matter is in a common envelope then, because the galaxies react mainly to the potential of the full group, the dense subgroups are usually unbound, as suggested by Rose (1979). The frequency of chance alignments increases with increasing number of galaxies and decreasing parent group size as predicted in Walke and Mamon's (1989) analytical study. Its global value is a few percent for standard (POSS E-band surface magnitude $\mu \leq 26$) HCGs, but reduced to typically 10^{-3} for very compact ($\mu < 24$) HCGs.

Table 2 shows that binaries occur often within 1D chance alignments. The fraction of chance alignments with one single binary is 50%, two binaries: 10%,

TABLE 1. Compact Configurations within Simulated Loose Groups

Model	Dark Matter	P(CA)	3×3D/1D	bound 3D/all 3D
Standard	Halos	1.1%	7.8%	81%
$R = 500\,h^{-1}$ kpc	Halos	2.0%	12.4%	90%
$N = 12$	Halos	3.3%	4.3%	86%
$(M/L)_{\rm gp} = 200\,h$	Halos	1.4%	14.5%	98%
Standard	Envelope	4.8%	13.9%	33%
$R = 500\,h^{-1}$ kpc	Envelope	6.5%	19.0%	22%
$N = 12$	Envelope	10.3%	8.0%	26%
$(M/L)_{\rm gp} = 200\,h$	Envelope	5.7%	12.3%	43%

Notes: $P(CA)$ is the frequency of chance alignments. The next column gives the fraction of chance alignments that are 3 dimensional. The final column lists the fraction of 3D subgroups that are bound.

TABLE 2. Chance Alignments along the Viewing Axis

Dark Matter	1+1+1+1	2B+1+1	2B+2B	2U+1+1	2U+2U	3B+1	3U+1
Halos	28%	50%	9%	<1%	<1%	11%	1%
Envelope	26%	12%	1%	39%	8%	4%	10%

Note: B and U stand for bound and unbound, respectively.

a triplet: 10 to 15%. *Less than 30% of all chance alignments have no binaries or higher multiplicity systems* (these are the 1+1+1+1 configurations). The numbers above depend very little on the location of the dark matter within the parent loose groups. However, if the dark matter lies in galaxy halos, most of these binaries and triplets are bound, and conversely if the dark matter is in a common envelope most of the binaries and triplets are unbound.

One expects that bound binaries and triplets produce slow and hence strong dynamical interactions, while unbound binaries and triplets produce fast and weak interactions. Considering then both 1D and 3D subsystems that appear like HCGs in the simulated loose groups, using Tables 1 and 2, one finds that the fraction of groups with at least 3 interacting galaxies is 19% or 27% depending on whether the dark matter is in individual halos or in a common envelope. Similarly, the fraction of interacting galaxies is either 47.5% or 50.0%, respectively. These percentages agree with the 12% to 33% of HCGs having at least 3 morphologically disturbed galaxies (see §3.1) and the 33% (morphology — §3.1) to 54% (spiral kinematics — §3.5) of galaxies showing signs of interaction. Therefore, *roughly half of the galaxies in simulated Hickson compact groups would show signs of dynamical interaction, when only $\simeq 10\%$ of these groups are "real".*

4.3. Alternatives

Among the alternative hypotheses listed above, some can be ruled out quite easily to explain the majority of HCGs, although they may well explain a few of these groups.

Spectroscopic measurements have been performed on all the galaxies in HCGs and, although these have still not been published, the latest published subset (Hickson, Kindl and Huchra 1988; Tikhonov 1990) suggests that perhaps 10% of the HCGs would no longer be quartets if one removed the galaxies with grossly discordant redshifts.

So far, only two HCGs (HCG 18 and HCG 54) seem to have been misclassified single galaxies whose different component may correspond to HII regions (Williams and van Gorkom 1988; Tikhonov 1990).

If HCGs were loose groups at cosmological collapse, then their mass-to-light ratios ought to be double those of virialized loose groups, since their kinetic energy will be roughly equal to the absolute value of its potential energy instead of half of that in virial equilibrium. Instead, the mass-to-light ratios of HCGs are roughly one-third of the corresponding loose group M/Ls (§4.1), although it remains to be verified what fraction of these loose groups are indeed virialized.

Finally, many properties of HCGs are very distant dependent (Tikhonov 1990; Mamon 1990; Whitmore 1990). The most distant HCGs tend to be elliptical rich, luminous, large, and with higher velocity dispersions. All these trends are features associated with rich clusters of galaxies. I have therefore suggested (Mamon 1993) that a non-negligible number of the distant HCGs represent the bright-ends of such clusters, which explains the trends observed by Rood and Williams (1989) and also reported by Palumbo (1993) in the environments of HCGs: the elliptical fraction of HCGs is higher than its surroundings when these are rich, whereas HCGs in poor environments have the same elliptical fraction as their environments.

5. SUMMARY

The compact groups in Hickson's catalog arise from a variety of origins. Statistical arguments favor *most* (but certainly not all) HCGs as arising from chance alignments in loose groups and clusters (Mamon 1986; Walke and Mamon 1989). These chance alignments tend to be binary-rich, thus explaining that *HCG galaxies show more signs of dynamical interaction and star formation than field galaxies, but less than binary galaxies* (see Rubin et al. (1991) and Moles (1992), respectively).

Table 3 below gives an estimated rundown of the different hypotheses. Notice the total of 100 HCGs. These values have been obtained as follows: ¿From the recent N-body simulations there should be 8 times more chance alignments in groups or clusters than bound or unbound systems forming within loose groups and clusters. As the distribution of dark matter in loose groups is still unknown, the average between the two probabilities for bound 3D systems is taken, giving 57% of bound systems among the 3D subgroups in the standard runs (see Table

TABLE 3. Estimate of the Frequencies of Hickson Compact Group Origins

Nature	Number
Primordial Bound Dense Groups	5
Recently Formed Bound Dense Groups	5
Unbound Dense Subgroups of Loose Groups	3
Chance Alignments in the Field (incl. Triplets)	10
Chance Alignments within Loose Groups	40
Chance Alignments within Clusters	20
Collapsing Loose Groups	5
Bright-Ends of Clusters	10
HII Regions	2

1). Then altogether, the difficulty in forming dense groups and the statistical arguments on group M/L, $\langle m_1 - m_2 \rangle$, and morphology-density relation seem to impose that the majority of HCGs are not real. The values in Table 3 then follow, assuming a rather optimistic value of 5 primordial dense groups, and a wild guess for the number of collapsing groups (which cannot be too large or else so would be M/L), and with the observational constraints on alternative hypotheses mentioned in §4.3.

6. REFERENCES

Allen, R.J., and Hartsuiker, J.W. 1972, *Nature*, **239**, 324.
Bahcall, N.A., Harris, D.E., and Rood, H.J. 1984, *ApJ*, **284**, L29.
Baier, F.W., and Tiersch, H. 1979, *Astrofiz.*, **15**, 33.
Bender, R., Döbereiner, S., and Möllenhoff, C. 1988, *A&AS*, **74**, 385.
Biermann, P., Kronberg, P.P., and Madore, B.F. 1982, *ApJ*, **256**, L37.
Binney, J.J., and Petrou, M. 1985, *MNRAS*, **214**, 449.
Burstein, D., Davies, R.L., Dressler, A., Faber, S.M., Stone, R.P.S., Lynden-Bell, D., Terlevich, R.J., and Wegner, G. 1987, *ApJS*, **64**, 601.
Djorgovski, S., Weir, N., and de Carvalho, R.R. 1992, in *Cosmology and Large Scale Structure in the Universe*, ASP Conference Series 24, ed. R.R. de Carvalho, San Fransisco, A.S.P., p. 141.
Efstathiou, G., Ellis, R.S., and Peterson, B.A. 1988, *MNRAS*, **232**, 431.
Faber, S.M., and Jackson, R.E. 1976, *ApJ*, **204**, 668.
Governato, F., Bhatia, R and Chincarini, G 1991, *ApJ*, **371**, L15.
Heiligman, G.M., and Turner, E.L. 1980, *ApJ*, **236**, 745.
Hickson, P. 1982, *ApJ*, **255**, 382.
Hickson, P. 1990, IAU Colloquium 124, *Paired and Interacting Galaxies*, eds. J.W. Sulentic and W.C. Keel, Washington, NASA, p. 77.
Hickson, P., Kindl, E., and Auman, J.R. 1989, *ApJS*, **70**, 687.
Hickson, P., Kindl, E., and Huchra, J.P. 1988, *ApJ*, **331**, 64.
Hickson, P., Menon, T.K., Palumbo, G.G.C., and Persic, M. 1989, *ApJ*, **341**, 679.

Hickson, P., Ninkov, Z., Huchra, J.P., and Mamon, G.A. 1984, in *Clusters and Groups of Galaxies*, eds. F. Mardirossian, G. Giuricin and M. Mezzetti, Dordrecht, Reidel, p. 367.
Hickson, P., and Rood, H.J. 1988, *ApJ*, **331**, L69.
Loveday, J., Peterson, B.A., Efstathiou, G., and Maddox, S.J. 1992, *ApJ*, **390**, 338.
Mamon, G.A. 1986, *ApJ*, **307**, 426.
Mamon, G.A. 1987, *ApJ*, **321**, 622.
Mamon, G.A. 1990, in *Paired and Interacting Galaxies*, IAU Colloquium 124, eds. J.W. Sulentic and W.C. Keel, Washington, NASA, p. 619.
Mamon, G.A. 1992, *ApJ*, in press.
Mamon, G.A. 1993, in *Distribution of Matter in the Universe*, 2nd DAEC meeting, eds. G. Mamon and D. Gerbal, Paris, Obs. de Paris, p. 51.
Mendes de Oliveira, C., and Hickson, P. 1991, *ApJ*, **380**, 30.
Menon, T.K. 1990, in Dynamics and Interactions of Galaxies, ed. R. Wielen, Berlin, Springer, p. 423.
Menon, T.K., and Hickson, P. 1985, *ApJ*, **296**, 60.
Moles, M. 1992, these proceedings.
Palumbo, G.G.C. 1993, in *Distribution of Matter in the Universe*, 2nd DAEC meeting, eds. G. Mamon and D. Gerbal, Paris, Obs. de Paris, p. 46.
Perea, J. 1992, these proceedings.
Postman, M., and Geller, M.J. 1984, *ApJ*, **281**, 95.
Ramella, M., Geller, M.J., and Huchra, J.P. 1989, *ApJ*, **344**, 57.
Rood, H.J., and Williams, B.A. 1989, *ApJ*, **339**, 772.
Rose, J.A. 1977, *ApJ*, **211**, 311.
Rose, J.A. 1979, *ApJ*, **231**, 10.
Rubin, V.C., Hunter, D., and Ford, W.K., Jr. 1991, *ApJS*, **76**, 153.
Shakhbazyan, R.K. 1973, *Astrofiz.*, **9**, 495.
Sulentic, J.W. 1987, *ApJ*, **322**, 605.
Sulentic, J.W. 1992, these proceedings.
Sulentic, J.W., and Mello Rabaça, D. 1992, *ApJ*, in print.
Thuan, T.X. 1992, in *Physics of Nearby Galaxies: Nature or Nurture?*, XIIth Moriond Astrophysics Meeting, eds. T.X. Thuan, C. Balkowski and J. Trân Thanh Vân, Gif-sur-Yvette, eds. Frontières.
Tikhonov, N.A. 1990, in *Paired and Interacting Galaxies*, IAU Colloquium 124, eds. J.W. Sulentic and W.C. Keel, Washington, NASA, p. 105.
Toomre A. 1977, in *The Evolution of Galaxies and their Populations*, eds. B.M. Tinsley and R.B. Larson, New Haven, Yale University Observatory, p. 401.
Tremaine, S.D., and Richstone, D.O. 1977, *ApJ*, **212**, 311.
Tully, R.B. 1987, *ApJ*, **321**, 280.
Turner, E.L., and Gott, J.R. 1977, *ApJS*, **32**, 409.
van der Hulst, J.M., and Rots, A.H. 1981, *AJ*, **86**, 1775.
van Gorkom, J.H., and Williams, B.A. 1992, private communication.
Walke, D.G., and Mamon, G.A. 1989, *A&A*, **295**, 291.
Williams, B.A. 1992, these proceedings.
Williams, B.A., McMahon, P.M. and van Gorkom, J.H. 1991, *AJ*, **101**, 1957.
Williams, B.A., and Rood, H.J. 1987, *ApJS*, **63**, 265.
Williams, B.A., and van Gorkom, J.H. 1988, *AJ*, **95**, 352.
Zepf, S.E. 1992, these proceedings.
Zepf, S.E., and Whitmore, B.C. 1991, *ApJ*, **383**, 542.
Zepf, S.E., Whitmore, B.C., and Levison, H.F. 1991, *ApJ*, **383**, 524.

Dynamics of Early-Type Galaxies in Hickson Compact Groups[1]

D. BETTONI AND L.M. BUSON
Osservatorio Astronomico di Padova, Vicolo dell'Osservatorio 5, I-35122 Padova, Italy

L. MAIRA AND F. BERTOLA
Dipartimento di Astronomia, Università di Padova, Vicolo dell'Osservatorio 5, I-35122 Padova, Italy

ABSTRACT: In this paper we present and discuss kinematical data obtained for 22 Elliptical and S0 galaxies belonging to 14 Hickson compact groups. One possible case of counter-rotation of gas and stars is discussed.

1. INTRODUCTION

Hickson groups are small associations of galaxies apparently in close proximity (Hickson 1982). The criteria defining a group are the number of objects, their isolation and the compactness of the group itself.

The main interest in the study of compact groups comes from the morphological and spectroscopic evidence showing that they are generally dynamical entities. In consequence of their properties, namely short crossing times, small velocity dispersions and high space densities, mergers and tidal effects should be favoured. Moreover, unlike cluster galaxies where intergalactic matter can be efficient in sweeping out the gas, the material possibly acquired by these galaxies could more easily reach and retain a stable configuration.

In particular, one can expect cases where the kinematical major axis of the gas does not coincide with the kinematical major axis of the stellar body of the galaxy (Dcmoulin Ulrich et al. 1984, Bertola et al. 1984, Bertola and Bettoni, 1988, Bertola et al. 1992). The more convincing explanation of such a kinematical decoupling is that the observed gas comes from a second event in the history of the galaxy, being the result of acquisition from outside (e.g., Bertola et al. 1988).

This is confirmed by Rubin (1988) who studied a sample of Hickson groups. She found that most of the galaxies, though relatively normal from the morphological point of view, have rotation patterns which are moderately abnormal. In addition, virtually all of the observed E and S0 galaxies contain small nuclear gas disks. In the case of the S0 galaxy H23c = NGC 1216 the gas disk appears counter-rotating with respect to the stellar body, showing that the stellar and gaseous components are kinematically decoupled.

---

[1] Based on observations obtained at the European Southern Observatory, La Silla, Chile

2. OBSERVATIONS AND DATA REDUCTION

Spectra of early–type galaxies belonging to nine southern Hickson groups were obtained with the ESO 1.52 m telescope equipped with the B&C spectrograph on September 21–23 1990 and May 11–13 1991. In the two consecutive runs, a 600 grooves mm^{-1} grating (giving a dispersion of 59 Å mm^{-1} at the 2nd order) was combined with a 1024 × 640 pixel RCA CCD and a 1024 × 1024 pixel Thomson CCD respectively, yielding a wavelength range $\sim \lambda\lambda$ 4600–5900 Å with a spectral resolution of $\sim$ 2 Å. Galaxies of five additional groups were observed in September 1991 and February 1992 with the Asiago 1.82 m telescope equipped with the B&C spectrograph and a 580 × 380 pixel Thomson CCD detector. The adopted grating gives a dispersion of 85 Å mm^{-1} with a spectral resolution of $\sim$ 4 Å in the wavelength range $\lambda\lambda$ 5000–6000 Å.

Each galaxy was observed at several position angles. The chosen wavelength region includes both the [O III] emission line (λ 5007 Å) and the absorption lines Mg I (λ 5175 Å), E-band (λ 5269 Å) and Fe I (λ 5335 Å). The exposure time ranged from 60 to 120 minutes. On each night, at least two spectra of slowly-rotating giant stars of spectral type from late-G to early-K were also taken as templates for the Fourier quotient reduction.

Spectra were reduced in a standard way using the IRAF[2] package. The Fourier Quotient technique (Bertola et al. 1984) was applied to all spectra in order to obtain the stellar velocity and velocity dispersion curves. The gas kinematics was analysed by fitting gaussian profiles to the emission lines with a package developed by G. Galletta.

3. RESULTS

In Table 1 we summarize the main properties of the galaxies included in our sample. Object name, morphological type, extinction corrected B_T magnitude and apparent ellipticity ϵ taken from Hickson et al. (1989) are given in columns (1)–(4), respectively. Column (5) gives the radial velocity corrected to the Galactic Standard of Rest taken from the RC3 catalogue (deVaucouleurs et al. 1991) or the PGC catalogue (Paturel et al. 1989). Columns (6)–(8) contain the observed maximum rotation velocity V_{max}, the central velocity dispersion σ_c (averaged over the central 3 arcsec) and the radius r_{max} at which V_{max} is reached.

The majority of galaxies belonging to our sample are giant ellipticals. In the (V_{max}/σ_c, ϵ) plane they appear as slow rotators, as expected for this kind of objects. On the contrary, S0 galaxies are mostly fast rotators. This result reflects the usual properties of early-type galaxies.

The dynamical behaviour of these objects appears possibly peculiar in the relationship between central velocity dispersion σ and absolute magnitude M_B, in the sense that the central velocity dispersion is sistematically lower than the classical value predicted for elliptical galaxies of the same luminosity (Faber and Jackson 1976). Compact group ellipticals included in our sample and those

---

[2]IRAF is distributed by NOAO, which is operated by the AURA, Inc., under cooperative agreement with the NSF, USA.

studied in literature (Dressler et al. 1987 and Lauirikainen 1990) seem to be in better agreement with the relation (L $\propto \sigma^{5.4}$) proposed by Kormendy and Illingworth (1983).

TABLE 1

Hickson #	Type	B_{TC}	ϵ	v_{GSR} km s^{-1}	V_{max} km s^{-1}	σ_c km s^{-1}	r_{max} arcsec
H14B	E5	14.2	0.46	5356	70	191±64	3.5
H14B GAS					60		2
H22A	E2	12.2	0.22	2616	100	214±14	6
H22A GAS					120	83±26	5
H37A	E7	13.0	0.83	6648		250±30	
H37A GAS					150	242±37	7
H48A	E2	13.2	0.23	2803	60	306±10	5
H48C	S0a	15.8	0.56	3999	100	119±22	5
H48D	E1	16.7	0.15	2841	50	99±99	2.5
H62A	E3	13.4	0.34	4249	70	272±10	5
H62B	S0	13.8	0.25	3545	160	290±117	6
H65B	S0	14.5	0.57	14576	110	238±5	4
H65C	E2	14.8	0.22	14000	70	255±25	4
H67A	E1	12.7	0.08	7191	120	300±31	2.5
H67D	S0	15.3	0.41	7005	80	151±36	2.5
H67D GAS					70		3
H68A	S0	11.8	0.23	2187	340	303±13	17
H68A GAS					180	208±23	3
H68B	E2	12.2	0.18	2539	50	226±6	3.5
H68B GAS					40		3
H86A	E2	13.7	0.2	6215	150	318±39	2.5
H86B	E2	14.2	0.25	6237	140	234±5	3.5
H86C	SB0	15.1	0.23	5569	100	186±34	4
H86D	S0	15.0	0.49	5957	75	214±45	2
H90B	E0	12.6	0.09	2559	50	259±11	2.5
H90C	E0	12.7	0.17	2535	50	211±10	3
H93A	E1	12.6	0.08	5278		240±18	
H98A	SB0	13.7	0.52	7938	200	258±31	5

The measured stellar rotation curves are generally confined to the very central region and do not show heavy distortions. In some cases they can be tracked to larger distances on the side of a close companion where the velocity appears to become higher.

The observed stellar velocity dispersion profiles do not differ from the usual profiles – constant or decreasing with radius – of other early-type galaxies. However, the presence of a nearby galaxy seems to induce a rise of the velocity dispersion on the side towards the companion galaxy.

Only in 6 (4 E and 2 S0) out of 22 E and S0 galaxies we observed, emission lines have been detected. This detection rate is not higher than the usual fraction of early–type galaxies showing ionized gas in their central regions (e.g., Roberts

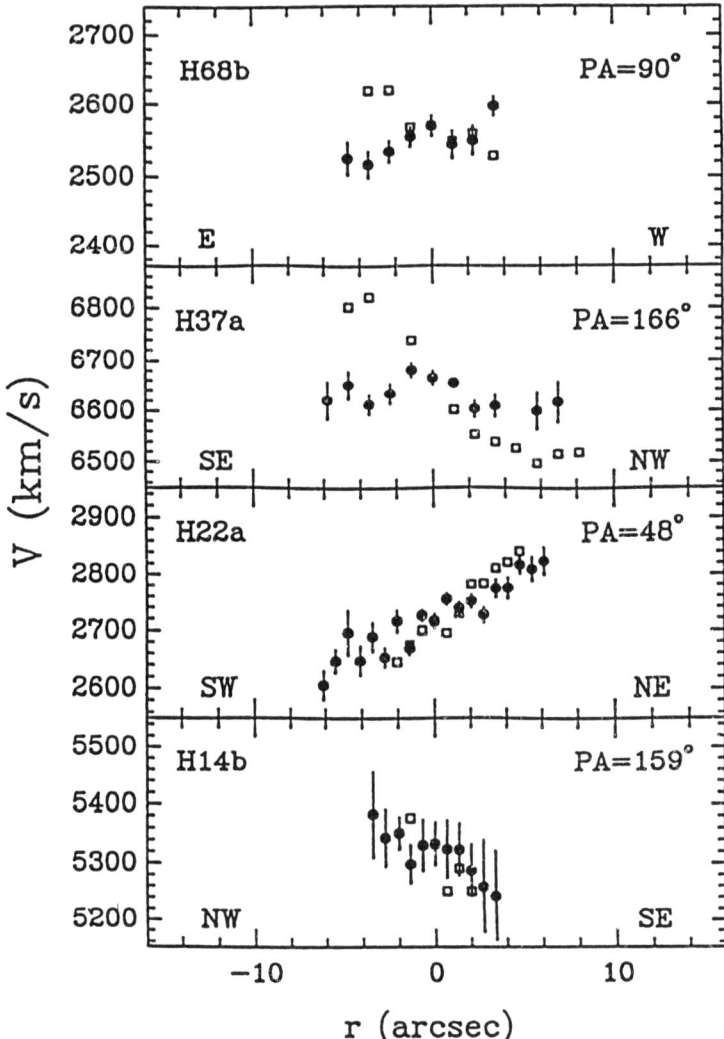

Figure 1. Major-axis gas and stellar rotation curves for elliptical galaxies with observed emission lines. Filled circles and open squares represent stars and gas respectively.

et al. 1991). In Fig. 1 the derived rotation curves of gas and stars for ellipticals are given. Two objects, namely H14b and H22a, show gas and stars rotating with a similar gradient. Gas and stars appear to rotate with different velocities in H37a, where a flat stellar rotation curve coexists with a strong velocity gradient of the gas. H68b deserves a special attention, being apparently an elliptical whose gas and stellar components are counter-rotating. The two rotation curves show clearly an opposite slope in the inner 8 arcsec, thus suggesting that the angular momentum vectors of the two components are strongly decoupled.

4. REFERENCES

Bertola, F., Bettoni, D., Rusconi, L., and Sedmak, G. 1984, *AJ*, **89**, 356.
Bertola, F. and Bettoni, D. 1988, *ApJ*, **329**, 79.
Bertola, F., Buson, L., and Zeilinger, W.W. 1988, *Nature*, **335**, 705.
Bertola, F., Buson L.M., and Zeilinger W.W. 1992, *ApJ*, **401**, L79.
Demoulin-Ulrich, M.-H., Butcher, H.R., and Boksenberg, A. 1984, *ApJ*, **285**, 527.
Dressler, A., Lynden-Bell, D., Burstein, D., Davies, R.L., Faber, S.M, Terlevich, R.J., and Wegner, G. 1987, *ApJ*, **313**, 42.
deVaucouleurs, G., deVaucouleurs, A., Corwin, H.G., Jr., Buta, R.J., Paturel, G., and Fouqué, P. 1991, *Third Reference Catalogue of Bright Galaxies*, Springer-Verlag: New York.
Faber, S.M., and Jackson, R.E. 1976, *ApJ*, **204**, 668.
Hickson, P. 1982, *ApJ*, **255**, 382.
Hickson, P., Kindl, E., and Auman, J.R. 1989, *ApJS*, **70**, 687.
Kormendy, J., and Illingworth, G. 1983, *ApJ*, **265**, 632.
Lauirikainen, E. 1990, *A&A*, **232**, 323.
Paturel, G., Fouqué, P., Bottinelli, L., and Gouguenheim, L. 1989, *Catalogue of Principal Galaxies*, Observatoire de Lyon: France.
Roberts, M.S., Hogg, D.E., Bregman, J.N., Forman, W.R., and Jones, C. 1991, *ApJS*, **75**, 751.
Rubin, V.C. 1988, in *Large scale motions in the Universe: a Vatican study week*, p. 541.

Morphology of Early–Type Galaxies in Compact Groups

D. BETTONI AND G. FASANO
Padova Astronomical Observatory, Vicolo dell'Osservatorio 5,
I-35122 Padova, Italy

1. INTRODUCTION

Compact Groups of Galaxies (hereafter CGGs) offer a valuable sample in which to study the environmental effects on galaxy morphology. In general a galaxy group is defined as a CGGs when the mean projected separation among the galaxies belonging to the group is comparable to the diameters of the galaxies themselves and the group is sufficiently isolated. Several different lists of CGG have been proposed in the literature following different selection criteria. Among them we mention those produced by Shakhbazyan (1973), Petrosyan (1974, 1978), Rose (1977) and Hickson (1982), the first three being indeed lists of Compact Groups of Compact Galaxies.

CGGs are systems having small total mass but very high galaxy space density and short dynamical times (Hickson, Richstone and Turner 1987). Moreover, since typical relative velocities of galaxies in CGGs are much lower than those measured in clusters and comparable to the internal stellar velocities (Hickson and Huchra 1990), strong interactions in CGGs should be frequent. Actually, dynamical and spectrophotometrical studies (Rubin et al. 1991; Bettoni et al. 1992) and N–body simulations (Barnes 1989) suggest that galaxies in CGGs merge to form a single remnant, probably an elliptical galaxy. This is confirmed by the high percentage of early-type galaxies observed in CGGs (Hickson 1982; Hickson et al. 1989; Tikonov 1989), enforcing the hypothesis that elliptical galaxies are in general the product of interactions and merging processes (Barnes 1989, Nieto 1988).

A serious problem of this framework is represented by the observation that the group velocity dispersion increases at increasing the frequency of elliptical galaxies (Hickson et al. 1988) suggesting that the high percentage of early-type galaxies observed is most likely due to environmental effects at the time of galaxy formation.

Investigations on the morphology of early-type galaxies in compact groups should help us to understand if we are looking at objects which have jet experienced encounters and merging events. In particular, since some recent studies (Bender et al. 1988, Nieto 1988) indicate boxy isophotes in ellipticals as possible signatures of recent merging processes, it could be of some interest to investigate if the frequency of boxiness in elliptical galaxies belonging to compact groups is different from that observed in field galaxies. Moreover the presence of nuclear dust lanes and of possible features (and/or discontinuities) in the luminosity profiles (extended halos, tidal truncation, change of slope, etc..) could give informations on the past interactions suffered by the galaxies.

In order to perform this kind of analysis, we selected from the various available lists, the CGGs showing an high percentage of early type objects. In this paper, we present some results concerning the luminosity and geometrical profiles of early-type galaxies belonging to several Hickson Compact Groups (HCGs) and Rose Compact Groups (RCGs).

2. OBSERVATIONS AND DATA REDUCTION

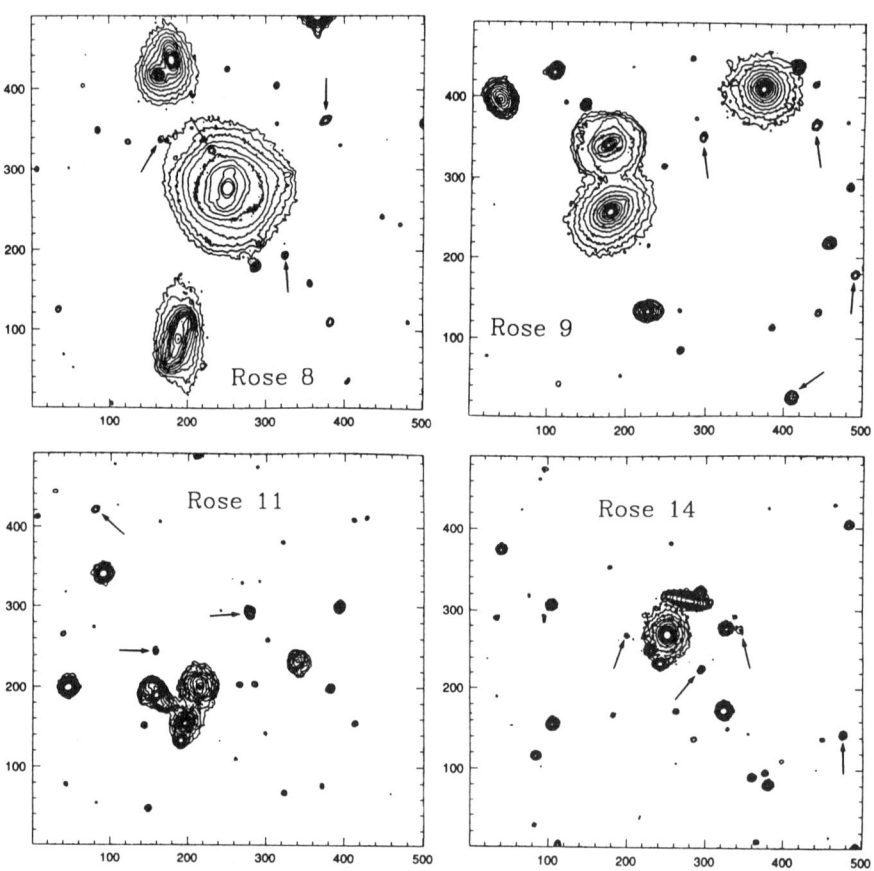

Figure 1. Isophotal contours of Rose CGs, obtained by sampling the surface brightness in steps of $0.^m5$.

The observations were carried out in the red band by using the 1.82 m. Asiago telescope equipped with the Tektronic TK512M CCD camera. A detailed description of the acquisition system is given in Bortoletto & D'Alessandro (1986). The CCD has a pixel format of 512×512, the $27\mu m$ square pixels giving a scale of $0.''34$ per pixel. Several flat-field frames were taken on the sky at the twilight of each night, while dark images have shown to be constant over the frame and almost negligible.

The preliminary reduction has been carried out in the canonical way, by using the Astronomical Images Analysis Package (Fasano 1990, hereafter AIAP) to perform flat fielding, bad pixels and/or columns removal and sky subtraction.

Luminosity and geometrical profiles have been also obtained by means of the AIAP software. This package provides the ellipse fitting of each isophote giving its major axis (a"), ellipticity (ϵ) and position angle (PA). In addition, it computes the well known a_4 coefficient, which has been widely used in the literature (see for example Bender et al. 1988) to quantify the deviations of the isophotal contour from the perfect ellipse ($a_4 > 0$ for the disky isophotes; $a_4 < 0$ for the boxy ones). The AIAP technique allows to mask interactively the isophotal frame, in order to eliminate parts of the image, whatever their shape may be. This capability is particularly useful in our sample, where there is an high frequency of interacting objects whose morphology is strongly perturbed. On the other hand, it is worth to stress that the fitting algorithm may give a_4 coefficients biased towards negative values when the isophotal contours are partially incomplete due to the masking procedure. For this reason we consider as reliable only the regions of the a_4 profiles where no relevant masking has been applied. In Figure 1 the isophotal contour of the frames of some Rose Compact Groups are shown. As noticed by Rubin et al. (1991), the presence of faint non–stellar objects (identifiable as small galaxies) in the fields of compact groups is rather common, in Figure 1 they will be marked by an arrow .

3. CONCLUSIONS

Our galaxy sample turns out to be composed by 47 ellipticals and 41 S0.

- The inner luminosity profile of almost all E and E/S0 galaxies, in our sample, is well represented by the de Vaucouleurs $r^{1/4}$ law, only few galaxies do not follow the $r^{1/4}$ law, and in these cases some peculiarities in the nuclear region (jet, double nucleus) are present. As for the outer luminosity profiles, we found that 40% of our ellipticals show an excess of light with respect to the de Vaucouleurs law, suggesting the presence of a tidal halo. The characteristic radius r_H at which the departure from the $r^{1/4}$ law become important is of the order of the effective radius ($r_e/r_H \sim 1.12$ in our sample). A similar result was found by Kormendy (1982), who noted also as the presence of tidal halos is more frequent in groups than in field and cluster ellipticals.

- Among the early–type galaxies studied, we find some objects with nuclear dust lanes. Our percentage (15%) of dust lane Es is lower than that found by Sadler and Gherard (1985). This difference can be due to the fact that our galaxy sample has a fainter average magnitude with respect to the Sadler and Gherard sample. We find also 3 cases of multiple nuclei (2 Es and 1 S0) and 2 cases of nuclear jets (1 S0 and 1 E/S0).

- Looking to the properties of the E and E/S0 galaxies belonging to our sample, we note a prevalence of the diskiness with respect to the boxiness, in fact only the $\sim$ 7% of our early–type galaxies shows boxy isophotes. This result seems rather different from that found by Bender et al. (1988) for a randomly selected sample in which almost the 50% of galaxies presents boxy isophotes. It is worth to stress also that the boxy ellipticals we found are quite apart from the central regions of the groups.

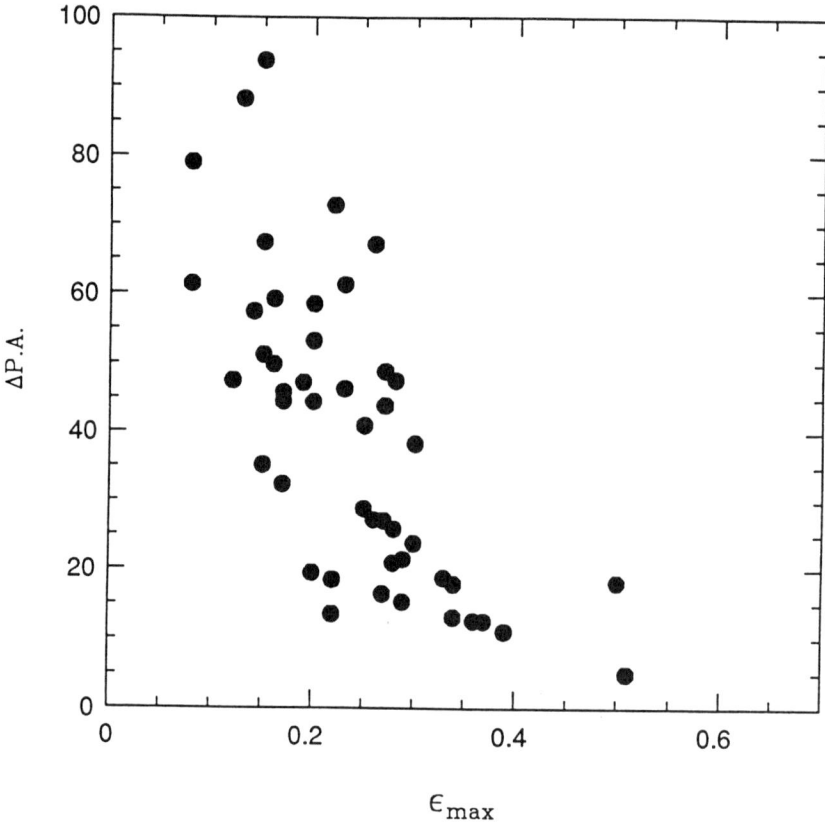

Figure 2. Correlation between maximum ellipticity and isophotal twisting for our sample of CGGs elliptical galaxies. We confirm Galletta's (1980) finding that a large twisting is found preferentially in low ellipticity galaxies.

- Figure 2 shows the relation between the maximum isophotal twisting $\Delta P.A.$ and the maximum ellipticity ϵ_{max} (Galletta 1980) for our sample of ellipticals. The fact that round objects have in general a larger isophotal twisting with respect to the flat ones is confirmed also in compact groups. Since Fasano and Bonoli (1989) and Rampazzo and Buson (1990) have shown that the above relation holds also for isolated and paired galaxies respectively, we conclude that its validity is independent on the environment. Figure 2 shows that in our sample only two elliptical galaxies are found with apparent ellipticity greater than 0.4. If confirmed, when a more complete database on early-type galaxies in CGGs will be available, this scarcity of objects in the queues of the ellipticity distribution will give a strong indication of triaxiality (Fasano and Vio 1991).

4. REFERENCES

Barnes, J. 1989, *Nature*, **338**, 123.
Bender, R., Dobereiner, S., and Möllenhoff, C. 1988, *A&AS*, **74**, 385.
Bettoni, D., Bertola, F., Buson, L.M., and Maira, L. 1992, in preparation.
Bortoletto, F., and D'Alessandro, M. 1986, *Rev.Sci.Instruments*, **57**, 253.
Fasano, G. 1990, Internal Report of the Padova Astronomical Observatory.
Fasano, G., and Bonoli, C. 1989, *A&AS*, **79**, 291.
Fasano, G., and Vio, R. 1991, *MNRAS*, **249**, 629.
Galletta, G. 1980, *A&A*, **81**, 179.
Hickson, P. 1982, *ApJ*, **255**, 382.
Hikson, P., Richstone, D.O., and Turner, E.L. 1977, *ApJ*, **213**, 323.
Hickson, P., Kindl, E., and Huchra, J.P. 1988, *ApJ*, **329**, L65.
Hickson, P., Kindl, E., and Auman, J.R. 1989, *ApJS*, **70**, 687.
Hickson, P., and Huchra, J.P. 1990, in preparation.
Kormendy, J. 1982, in *Morphology and dynamics of galaxies*, XX. Adv. Course Saas-Fe, p. 113.
Nieto, J.L. 1988, in *2da Reunion de Astronomia Extragalactica*, Academia Nacional de Ciencias de Cordoba, p. 239.
Petrosyan, M.B. 1974, *Astrophysics*, **10**, 471.
Petrosyan, M.B. 1978,*Astrophysics*, **14**, 631.
Rampazzo, R., and Buson, L.M. 1990, *A&A*, **236**, 25.
Rose, J.A. 1977, *ApJ*, **211**, 311.
Rubin, V.C., Hunter, D.A., and Ford, W.K. 1991, apjs, **76**, 153.
Sadler, E.M., and Gerhard, O.E. 1985, *MNRAS*, **214**, 177.
Shakhbazyan, R.K. 1973, *Astrophysics*, **9**, 495.
Tikonov, N.A. 1989, in IAU Coll. No. 124 *Paired and interacting Galaxies*, eds. J.W. Sulentic and C.M. Telesco, NASA Publ.No. 3098.

Star Formation and Merging in Hickson Compact Groups

MARIANO MOLES, ASCENSIÓN DEL OLMO, AND JAIME PEREA
Instituto de Astrofísica de Andalucía,
Apdo. 3004, 18080 Granada, Spain

ABSTRACT: Aperture photoelectric UBV photometry for 177 galaxies in Hickson groups is presented. Only moderate to small star formation rate enhancements are detected, a result confirmed by the analysis of IRAS data. On the other hand, merger candidates, as judged form the color indices, are very rare, if they exist at all. It is argued that the high density corresponding to compact groups could somewhat inhibit processes like those analysed here, as compared with situations of gravitational interactions in less dense environments.

1. INTRODUCTION

Physical entities like the compact groups of galaxies catalogued by Hickson (1982) pose a number of problems. Therefore, their reality has been seriously challenged, in particular by Mamon (1986). It is generally accepted, however, that at least a fraction of them are actually real groups, so in any case it seems sensible to analyze the primary consequences that would be expected. These are mainly an enhancement in the present star formation rate and the presence of galaxy mergers, often in the process of being formed.

The simplest way to characterize the present global star formation rate (SFR) in galaxies is through the use of UBV data. This was already shown by Larson and Tinsley (1978) who also stressed the independence of that indicator on other parameters like the internal reddening, the chemical composition, the adopted IMF or the functional form of the assumed SFR. The star formation activity can also be traced by the far infrared luminosity. Dultzin-Hacyan, Masegosa and Moles (1990) have shown that the best IR tracer is the luminosity at $25\mu m$ since those at longer wavelengths are more and more contaminated by the emission from dust heated by radiation from non ionizing stars.

The second aspect to consider when discussing high density regions is the merger frequency. It is expected that newly formed elliptical galaxies should be present in Hickson compact groups (HCGs). Their color indices would be bluer than those of old, well evolved normal E galaxies. However, it seems that the number of merger candidates found in recent studies (Zepf, Whitmore and Levison 1991; ZWL) is unexpectedly low. This result is confirmed and strengthened in this study.

2. OBSERVATIONS, DATA REDUCTION AND RESULTS

The data consist of UBV photoelectric photometry for 177 galaxies in 52 HCG, obtained with the spanish 1.52m telescope in Calar Alto (Almería, Spain), using a S20 EMI phototube running in counting mode. The counting errors were always kept under 1% except for the faintest objects in U for which they could become as large as 3%. Standard stars from the list by Neckel and Chini (1980) were observed nightly to determine the atmospheric extinction and the transformation coefficients of the instrumental magnitudes to the standard Johnson system. Taking into account all errors, our photometry is generally accurate at the 2% level with the exception of $(U - B)$ for the faintest objects for which the errors could rise to 4%.

Comparison of our data with those by ZWL (in all 27 galaxies in common) shows that the median value of the differences in (B-V) is 0.01 magnitudes, with a rms deviation amounting to 0.05 magnitudes. The situation regarding (U-B) is much less satisfactory. Excluding H49D for which the difference amounts to −0.31 magnitudes, the differences for the remaining points have a median value of 0.03 magnitudes, with a dispersion of 0.12 magnitudes. The most plausible reason for the discrepancy between our data and those by ZWL is the problem reported by ZWL in matching their instrumental system to the standard UBV system.

We also estimated the total B magnitudes, using the major axis length at $\mu_B = 25 \ mag/(")^2$ given by Hickson, Kindl and Aumann (1989; HKA) and the corresponding effective apertures, A_e. The comparison gives a zero point of 0.12 magnitudes (our values being fainter), and a scatter of 0.40 magnitudes. The comparison with data by ZWL gives no zero point difference and a dispersion of 0.2 magnitudes.

3. STAR FORMATION IN HICKSON GROUPS

The color-color diagrams for the observed galaxies grouped by morphological type are presented in Figures 1a to 1d, together with the locus of the average position for normal galaxies. It is worth noticing the scatter in the color-color diagram, which is much larger than expected from measurment errors alone. Moreover, the data points are not randomly distributed around the average position for normal galaxies, with the majority of the points located above that line. In other words, for a given $(B - V)_0$ value, the $(U - B)_0$ index tends to be bluer than for normal galaxies.

Those are clear indications for abnormal colors of galaxies in HCGs. That conclusion is similar to that found by Larson and Tinsley for the galaxies in Arp's Atlas (1966) and can be interpreted in terms of an enhancement in the present SFR with respect to the situation in normal galaxies. Those color changes are, however, only modest and indicate rather moderate star formation enhancements. Note that the most noticeable separations from the line for normal galaxies are for the earliest types, where the effects of even a small burst are visible against such a red background.

The preceding results are statistical in nature, but they nonetheless mean that none of the observed galaxies actually presents extreme blue or even peculiar colors as do the blue compact Zwicky galaxies, for instance (Moles et al. 1987). Furthermore, comparison with the color indices of galaxies in isolated pairs (Karachentsev 1972) with aperture photometry in the catalogue by Longo and de Vaucouleurs (1983) and well defined morphology, shows that the latter are even bluer than galaxies in HCGs.

Analysis of available IRAS data leads to similar conclusions. Hickson et

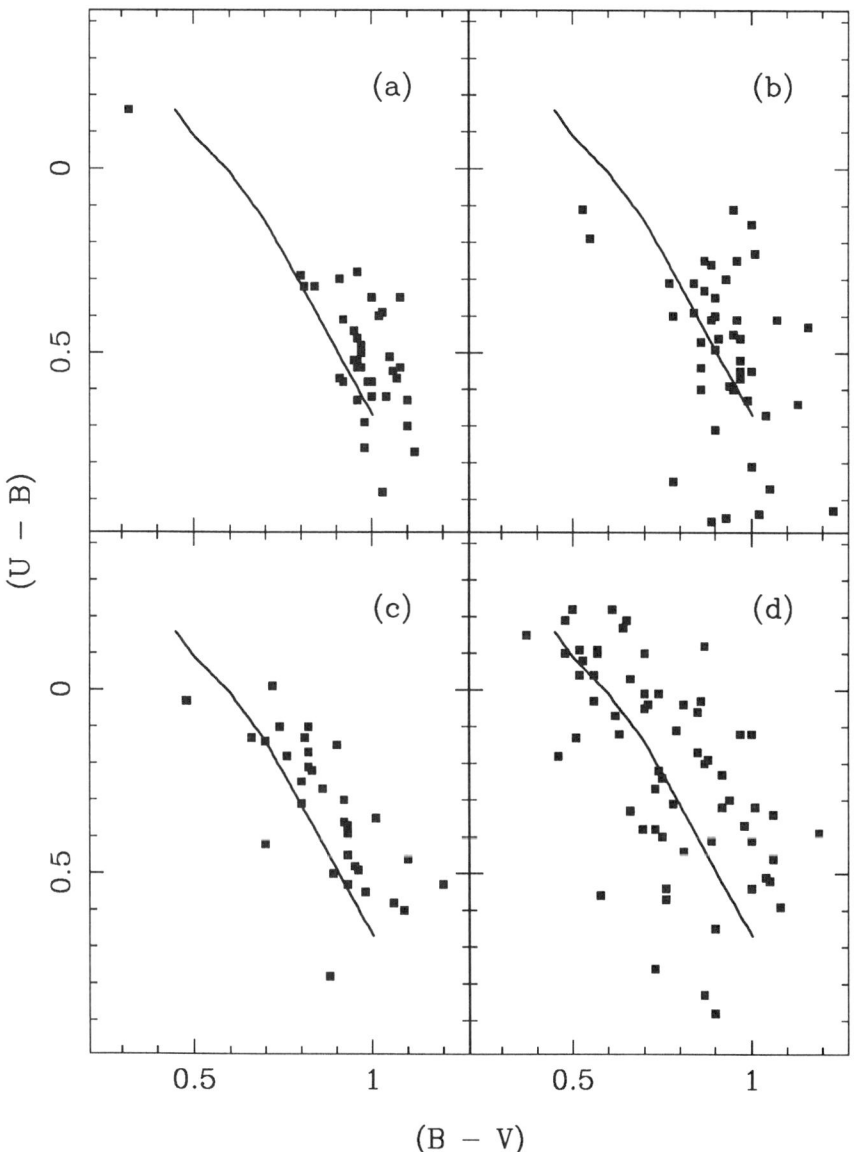

FIGURE 1. The color-color diagram for the observed galaxies: (a) Ellipticals, (b) S0-S0a, (c) Sa to Sb, (d) Sbc and later types.

al. (1989) considered the question and concluded from the analysis of the L_{FIR}/L_{opt} ratio that a sizeable enhancement of the SFR was detected. They estimated that about 1% of the high IRAS luminosity galaxies would belong to HCGs. Our results, taking only into account the actually detected cases (40 galaxies with reliable detections at 60 and 100μm) are far less clear. In Figure 2 it can easily see that the majority of the objects are inside the region occupied by normal, late type spirals. Leaving aside H63D, with a too low color temperature for its L_B/L_{FIR} ratio, and H96A (NGC 7674 = Arp182 = VV343 = Mkn533), a well known Seyfert galaxy (Laurikainen and Moles 1989), the remaining 10 objects (25% of the detected) could be genuine starburst galaxies. The point, however, is that only 3 of them were detected at 25μm, (H59D, H61C and H69B), out of the 9 that would have been detected in the case they are even normal emitters at that wavelength.

Another way to look for starburst objects is indeed to concentrate on the galaxies detected at 25μm (Dultzin-Hacyan, Masegosa and Moles 1990). Only 9 objects were certainly detected at 25 and 100μm. Let's note that, given the measured 100μm fluxes, and considering the I_{25}/I_{100} ratios found for normal galaxies (see Dultzin-Hacyan, Moles and Masegosa 1988) the number of detections at 25μm corresponds to what would be expected for normal spirals. This is again a clear indication about the rather small star formation enhancement among galaxies in HCGs.

The I_{25}/I_{100} ratio spans a range between 0.04 and 0.19, with a median value of 0.07 and a statistical dispersion of 0.05. As noted by Dultzin-Hacyan, Moles and Masegosa (1988), normal spiral galaxies have a median value of 0.04 and a dispersion of 0.02. Starburst and active objects appear to have ratios above 0.10. From the 9 objects detected at 25 and 100μm, only 4 have $I_{25}/I_{100} > 0.10$. Two of them, namely H59D (IC 737) and H69B were also detected as burst candidates from their position in the diagram in Figure 2. Thus, they meet all the characteristics to be the site of violently star forming processes. The third candidate is H31C, an irregular galaxy also catalogued as Mkn 1089. Its FIR to blue luminosity ratio is rather low, and its only outstanding property is its relatively large 25μm flux. Finally, the last candidate is H16C, also catalogued as Mkn 1022. No 60μm flux was detected for that object which belongs to a very interesting group. The analysis by Rubin, Hunter and Ford (1991) shows the dynamical and morphological peculiarities of the members of that group. In fact, three of the four member galaxies were detected by IRAS, even if only H16C was reliably detected at both 25 and 100μm.

The analysis of the IRAS data, together with that of the visible color indices, strongly suggest that even if the star formation in galaxies belonging to compact groups is enhanced with respect to that in normal, isolated spirals, it is not particularly intense. It could be not stronger than in galaxies in isolated pairs and, in any case, there are very few starburst objects in such aggregates, and all of them are irregular, low luminosity, blue galaxies.

4. MERGING IN HICKSON GROUPS

Under the merger hypothesis some, if not all the early type galaxies would be produced by the merging of two or more disk systems (Toomre and Toomre

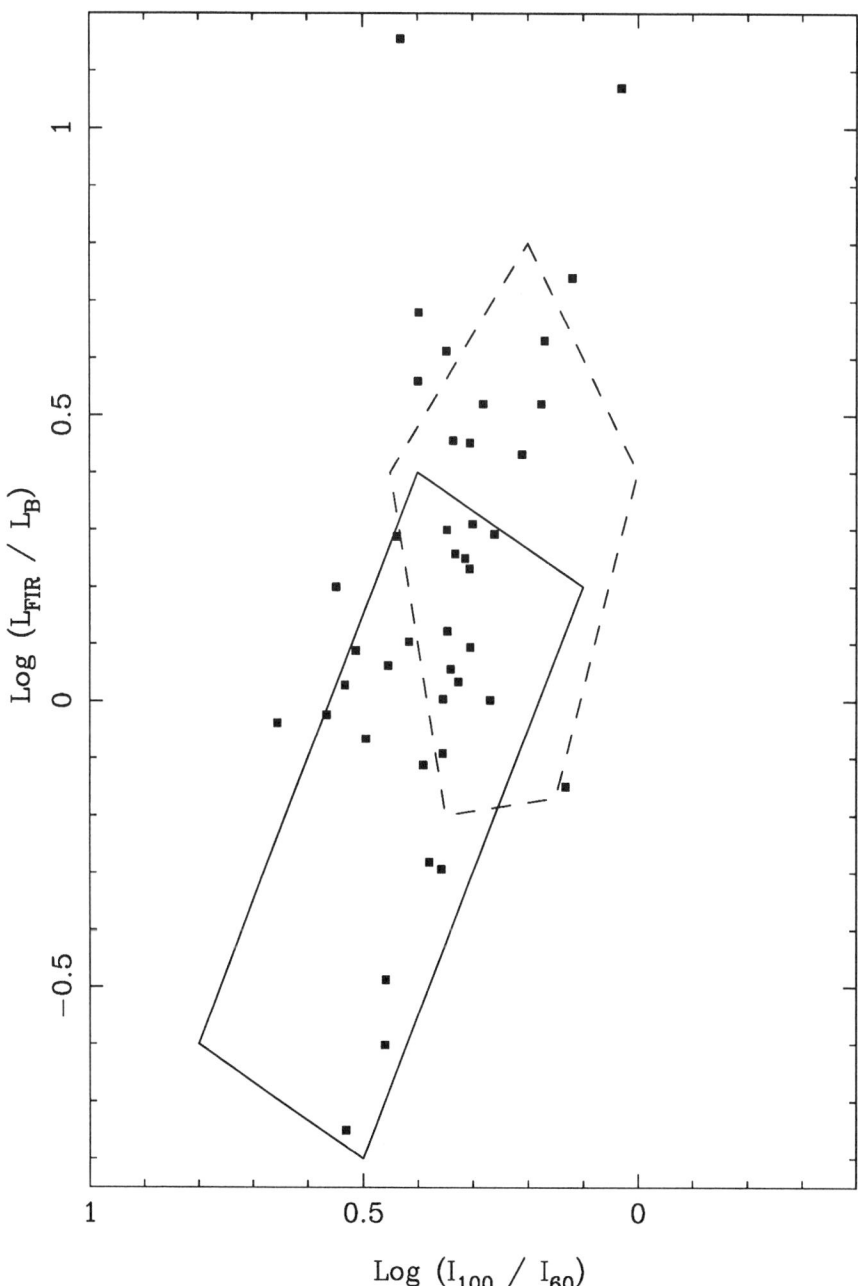

FIGURE 2. L_{FIR}/L_B versus I_{60}/I_{100} diagram for the galaxies detected by IRAS. The B luminosities have been calculated through the solar value. The solid line delineates the region occupied by normal spiral galaxies. The dashed line corresponds to the region occupied by the starburst galaxies.

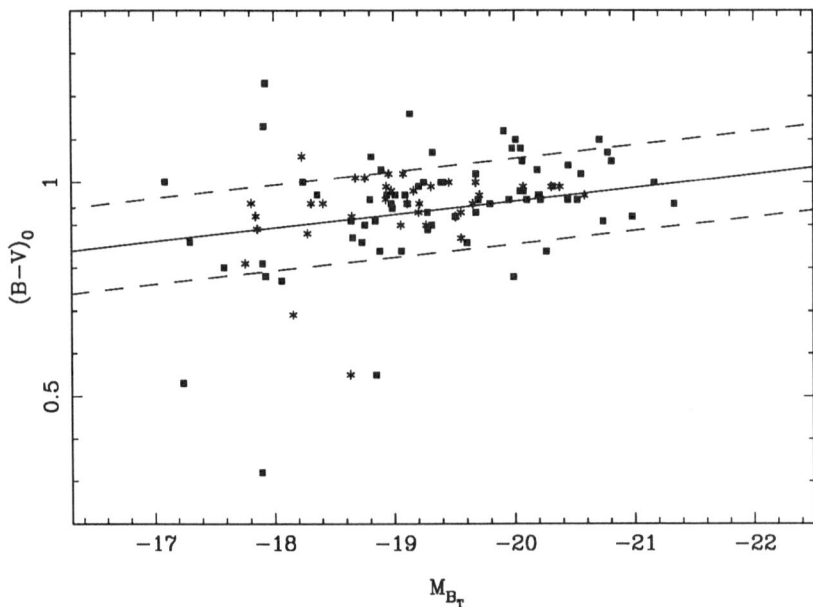

FIGURE 3. The color-magnitude diagram for early type galaxies in Hickson groups. Stars for data from ZWL, squares for our data. The solid line represent the mean relation for ellipticals from Burstein *et al.* 1987, and the dashed lines the 3σ parallels.

1972). The newly formed objects, when still young (age smaller than 1-2 Gyr), should appear as bluer than normal E galaxies. Indeed, numerical simulations of the dynamical evolution of galaxies in compact groups have shown that dynamical friction, tidal disruption and even galactic collisions can produce radical changes in the aspect of the group and of the member galaxies (Carnevalli, Cavaliere and Santangelo 1981; Barnes 1990). And all that in very short time scales corresponding to a few times 10^8yr (but see del Olmo 1988, and Governato, Bathia and Chincarini 1991 for results showing much longer merging scales). Then, genuine representatives of those stages should populate the HCGs, provided they are physical entities.

ZWL reported that only a few objects (at most) in HCGs would be candidates for mergers, none of them being a really strong case. Our study adds 49 galaxies to the sample analyzed by ZWL. The color-magnitude diagram for the whole set of data is presented in Figure 3, where the solid line gives the mean position of the ellipticals from the sample by Burstein *et al.* (1987). For the objects in common with ZWL we only plotted the values we find in the present work. It is clear that only a few objects can be considered as merger candidates. They are listed in table 1, where we give their most relevant properties. The candidates H4D and H56E, proposed by ZWL are included, but not H37A and H37E. These latter two galaxies are also contained in our sample and the apparent magnitudes are in excellent agreement with data by ZWL. However, the absolute magnitudes quoted by ZWL for them are not correct, probably because they didn't use the appropriate redshift value, z=0.0229.

None of the objects in table 1 is a first ranked galaxy, all being low luminosity, small objects. Moreover, considering both the (U-B) and (B-V) color indices they are practically indistinguishable from the old systems affected by newly formed stars (see Moles et al. 1987). Then, it can be said that the properties of those merger candidates are much better explained in terms of recent star formation events triggered by the accretion of gas directly form other galaxies in the group or from the intragroup medium.

TABLE 1. The photometric characteristics of merger candidates

Galaxy	$(B-V)_0$	$(U-B)_0$	$M_{B_{T_0}}$	D_e (kpc)
H4/D	0.55	0.09	−18.6	3.1
H8/D	0.84	0.31	−20.3	7.2
H49/D	0.32	−0.16	−17.9	5.8
H56/E	0.69	0.23	−18.2	6.3
H64/D	0.53	0.11	−17.2	1.6
H69/D	0.78	0.85	−17.9	2.6
H78/B	0.78	0.40	−20.0	5.3
H79/C	0.77	0.31	−18.1	4.3
H94/C	0.55	0.19	−18.9	6.0

5. ABOUT THE REALITY OF HCGS

The preceding results were unexpected in light of the hypothesis that HCGs are physically bound entities. The SFR enhancement is much lower than naively expected and the merging frequency seems to be nearly zero. Thus they could be taken as evidence against that hypothesis. It could then be argued, following Mamon, that the groups are only transient situations, chance superpositions of pairs and non related galaxies within loose groups.

In our opinion, however, the consequences of the results presented here are not quite so simple. First of all, the star formation in HCGs is definitively enhanced with respect to that in normal, locally isolated galaxies, but not as much as in isolated pairs. This is not easy to interpret in terms of chance alignments as proposed by Mamon. But, on the other hand, even the simulations by Mamon (see these proceedings) give about 20% of real groups, whereas the absence of starburst galaxies and mergers would imply, at its face value, that none of the HCGs is actually real. A conclusion difficult to sustain, indeed. Then, the explanation of the results reported here should be looked for in a different direction.

Let us recall that the expected star formation enhancement in HCGs is but an extrapolation from the analysis of some (particular) cases of interacting pairs and from model calculations. The behaviour of the molecular gas in such configurations is, however, not easy to trace. In that respect let us note that our results exclude galaxy-sized starbursts but not localized events within an object, in particular nuclear bursts. And it has been argued that to form a generalized burst in a galaxy some conditions are to be satisfied to inhibit partial fragmentation until the whole system experiences the burst (Campos-Aguilar and Moles 1992). In some sense, gravitationally hostile environments could inhibit such global processes, as indicated by the fact that the majority, if not all the Blue Compact Dwarf galaxies are locally isolated (Campos-Aguilar and Moles 1991).

Regarding the lack of merger candidates, which is another expectation from theoretical considerations and numerical simulations, the situation is again contradictory. It has been shown that some particular sets of initial conditions could produce much longer merging time scales (del Olmo 1988; Governato, Bathia and Chincarini 1991), and consequently reduce the expected number of mergers. The difficulty is perhaps in reducing that expectation to essentially zero.

TABLE 2. The velocity dispersions as a function of the morphological content

Morphological content	N	σ_V
Only Spirals	11	55
Spiral dominated	33	91
E+S0 dominated	24	240
Only E+S0	9	303

Finally, the general aspect of the groupings and their frequency (we have to consider also the Shakhbazyan groups, which are in many respects similar to the HCGs), strongly favors the view of their physical reality. And, as pointed out before, the problem is already there once it is accepted that even only some fraction of the groups are real. It seems then that we are forced to face the problems posed by the compact groups, whose properties seem to indicate that the consequences of the gravitational interactions in dense systems are different from what is seen in colliding pairs. Some of the observed properties, in particular the relation between morphological content and velocity dispersion (Hickson, Kindl and Huchra 1988; see table 2) could manifest themselves as fundamental to understand the nature of the groups and therefore should be taken into account in realistic simulations. After all, that tight relation would just indicate the class of systems that could survive in such extreme situations.

Thus, even if, as usually stated, more observations are needed to improve the situation, it seems that a full reanalysis of the situation on the theoretical side could also be very convenient to understand the basic aspects of the galaxies belonging to dense configurations.

6. REFERENCES

Arp, H.C. 1966, *Atlas of Peculiar Galaxies*, California Institute of Technology, Pasadena, USA.
Barnes, J. 1990, in *Dynamics and Interactions of Galaxies*, ed. R. Wielen, Springer-Verlag, Heidelberg, p. 186.
Burstein, D.C., Davies, R.L., Dressler, A., Faber, S.M., Stone, R.P.S., Lynden-Bell, D., Terlevich, R.J., and Wegner, G. 1987, *ApJS*, **64**, 601.
Campos-Aguilar A., and Moles, M. 1991, *A&A*, **241**, 388.
Campos-Aguilar A., and Moles, M. 1992, *ApJ*, in press.
Carnevalli, P., Cavaliere A., and Santangelo, P. 1981, *ApJ*, **249**, 449.
del Olmo, A. 1988, Tesis Doctoral, Universidad de Granada.
Dultzin-Hacyan, D., Moles, M., and Masegosa, J. 1988, *A&A*, **206**, 95.
Dultzin-Hacyan, D., Masegosa, J., and Moles, M. 1990, *A&A*, **238**, 28.
Governato, F., Bathia, R., and Chincarini, G. 1991, *ApJ*, **371**, L15.
Hickson, P. 1982, *ApJ*, **255**, 382.
Hickson, P., Kindl, E., and Huchra, J. 1988, *ApJ*, **331**, 64.
Hickson, P., Kindl, E., and Aumann, J.R. 1989, *ApJS*, **70**, 687.
Hickson, P., Menon, T.K., Palumbo, G.G.C., and Persic, M. 1989, *ApJ*, **341**, 679.
Karachentsev, I.D. 1972, *Soob.Sb.Astr.Observatory Akad. Nauk*, **7**.
Larson, R.B., and Tinsley, B.M. 1978, *ApJ*, **291**, 46.
Laurikainen, E. and Moles, M. 1989, *ApJ*, **345**, 176.
Longo, G., and deVaucouleurs, A. 1983, *Monographs in Astronomy* No. 3, Univ. of Texas, Austin, USA.
Mamon, G.A. 1986 *ApJ*, **307**, 426.
Moles, M., García-Pelayo, J.M., del Río, G., and Lahulla, F. 1987, *A&A*, **186**, 77.
Neckel, T.H., and Chini, R. 1980, *A&AS*, **39**, 411.
Rubin, V.C., Hunter, D., and Ford, W.K. 1991, *ApJS*, **76**, 153.
Toomre, A., and Toomre, J. 1972, *ApJ*, **178**, 623.
Zepf, S.E., Whitmore, B.C., and Levison, H.F. 1991, *ApJ*, **383**, 524.

The Shakbazyan Compact Groups and Their Populations

ASCENSIÓN DEL OLMO, MARIANO MOLES, AND JAIME PEREA
Instituto de Astrofísica de Andalucía,
Apdo. 3004, 18080 Granada, Spain

1. INTRODUCTION

Galaxy environment has prooven itself to be of fundamental relevance in the study of evolution of galaxies. Compact groups of galaxies with spatial densities even higher than those in the centers of rich clusters, albeit with lower velocity dispersions (thus favoring the encounters between galaxies), appear to be especially well suited sites to study the effects on the physical and structural properties of each member–galaxy, and to try to understand its connection with the dynamical state of the aggregate.

In addition to the well known catalogue of compact groups of galaxies created by Hickson (1982) there is a compilation of Compact Groups of Compact Galaxies initiated by Shakbazyan in 1957. She found on the Palomar Sky Survey plates a very compact group of red, compact objects, that was initially thought to be an association of stars. But in fact, as was later shown by Robinson and Wampler (1973), it is a group of galaxies at relatively high redshift (z=0.1168) and with a low velocity dispersion. This prompted a systematic search for compact groups which were similar to Shk 1. The adopted search criteria were the following:

i) Isolation and a minimum of 5 members.

ii) More than half of the galaxies in the group should appear compact in the red.

iii) The compact galaxies should appear very red.

A total of 377 groups were found that way (Shakbazyan 1973; Shakbazyan and Petrosyan 1974; Baier et al. 1974; Petrosyan 1974; Baier and Tiersch 1975, 1976a, 1976b, 1978; Petrosyan 1978; Baier and Tiersch 1979). At first they did not attract much attention. For several years the only available information was that contained in the identification publications, plus some photographic work (e.g. Arp et al. 1973; Börngen and Kalloglyan 1974; Ambartsumyan et al. 1975; Massey 1977; Shakbazyan 1978; Börngen and Kalloglyan 1980), and sparse spectroscopic data (Robinson and Wampler 1973; Mirzoyan et al. 1975; Kirshner and Malumunth 1980; Vennik et al. 1982). The few discussions were about the compactness of the galaxies and the rather low values found for the velocity dispersions.

We will see later that the Shakbazyan Compact Groups (ShCG) are similar to Hickson compact groups (HCG) in many respects, in particular in density and structure. But they present other specific aspects to be analyzed, as for instance the morphology and colors of the member galaxies. This will be discussed in the present contribution on the basis of new photometric and spectroscopic data for four groups (Shk 38, Shk 81, Shk 278 and Shk 362; see table 2 for their general properties). Three of them are well isolated, whereas Shk 81 is not and very probably lies at the center of a cluster of galaxies.

The data were collected with different telescopes at the Calar Alto and Roque de los Muchachos observatories in Spain. The details about the observations and the reduction procedures can be found in del Olmo (1988). They comprise both, CCD broad band images and spectroscopy.

2. THE POPULATION OF SHCG

The selection criteria were chosen to produce compact groups of red, compact galaxies. These are, therefore, the first properties to analyze. In fact, the work accumulated already earlier indicates that the galaxies are rather normal, bright early type objects (del Olmo 1988; Amirkhanyan et al. 1988; Kodaira et al. 1988, 1990; Lynds et al. 1990; del Olmo and Moles 1991). This suggestion will be corroborated here.

2.1. How compact are the galaxies?

We have used our surface photometry data to analyze first the contamination by stars and then the light distribution of the extended objects. In Figure 1 we present the profiles for the objects in the field of Shk 362 for illustration. The group is catalogued as containing six objects, numbers 1 to 6 in the figure. There are some other objects in the field, but it is clear that those with numbers 6, 7, 8, 9, 11 and 12 are in all probability stars. In particular, the stellar nature of # 6, previously catalogued as a galaxy, was confirmed spectroscopically. Thus the group is reduced to 5 bright objects plus eventually three faint objects. All these galaxies are well resolved. The same kind of analysis was performed for the other groups with analogous results.

Direct comparison of the observed profiles with those by Schombert (1986) clearly indicates that we are dealing with early type galaxies, generally of high luminosity (see Figure 2). This, together with the preceding results, means that, with the exception of object # 4 in Shk 278, which was meanwhile found to be a quasar (del Olmo and Moles 1991), the galaxies in Shk groups are not really compact but normal, early type objects. It now appears that they were incorrectly taken for compact in the search on the POSS because they are very luminous and distant (see below).

Another way to look into the question of morphology is through the relation between R_{24} and R_{22} worked out by Schombert. The data for the galaxies in our sample are plotted in Figure 3, where it can be seen that they are nicely distributed around the standard relation for normal early type objects. There are however two other aspects to notice in that figure. First of all, our objects tend to be above the standard line, which indicates that our galaxies are giant rather than normal ellipticals. Secondly, the objects belonging to Shk 81 populate the low end of the relation and depart form it. This agrees with the fact that the fraction of spirals in that group is higher, a fact that could be related to the non-isolation of that group.

Finally, the isophotal analysis also shows that they are early type objects, well fitted by the de Vaucouleurs $r^{1/4}$ law. Distortions are frequent but without

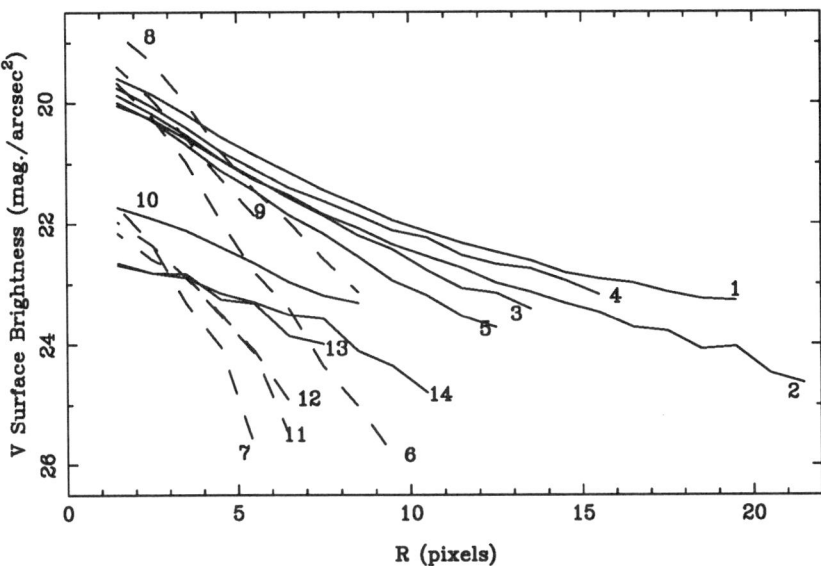

FIGURE 1. Observed profiles for the objects detected in the field of Shk 362

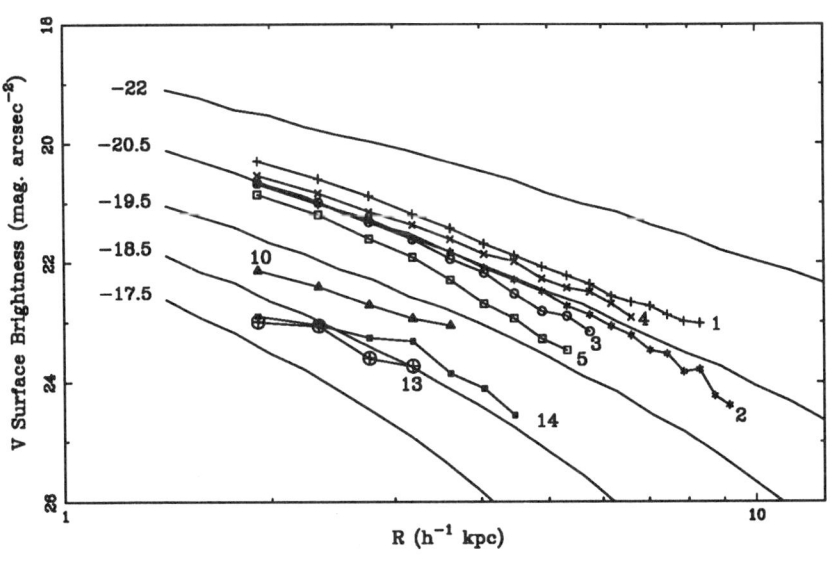

FIGURE 2. Photometric profiles in V for galaxies in Shk 362. The lines correspond to the Schombert template profiles for elliptical galaxies of different luminosities.

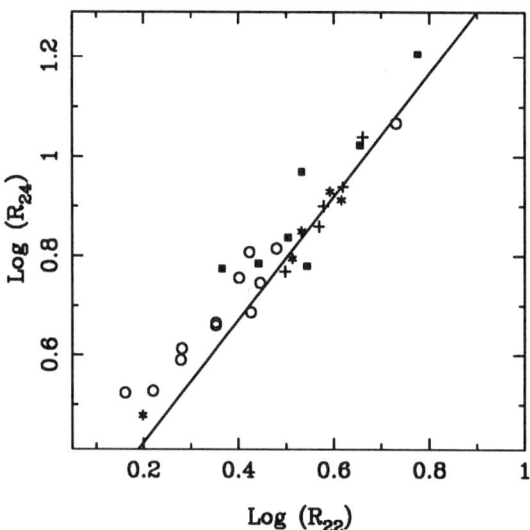

FIGURE 3. Diagram of the radii at the 22nd and the 24th isophotes in V band. + symbols for galaxies in Shk 362, * for Shk 38, o for Shk 81 and • for Shk 278.

any marked trends. In fact, the fourth coefficient, A_4, in the Fourier expansion of the residuals with respect to pure ellipses is significantly greater than zero in only a fraction of the cases, and almost never smaller than zero. In summary, almost 100% of the galaxies in the sample are early type objects. We found only three spirals, all of them in Shk 81. A fraction of 68% are ellipticals, and the other 30% are early types with indication of a faint disk as judged from the value of A_4. This is, of course, a big difference with respect to the situation in HCGs where about half of the members are spirals.

2.2. How red are the galaxies?

The observed colors are very red, indeed. But before jumping to conclusions, the corrections for the galactic extinction and the K− effect (which, at the redshift of the groups, is going to be important) have to be applied. The first was evaluated from the maps by Burstein and Heiles (1982); the K correction was estimated for the B and V bands, from the work by Pence (1976). The correction for the R and I bands was done just assuming a flat energy distribution in the range $\lambda\lambda 5500-10000\text{Å}$. For the galaxies for which we have spectroscopic data we could directly measure the amount of the K correction, finding that Pence's results are totally satisfactory. The corrected colors of the galaxies are plotted in Figure 4. The results clearly show that the galaxies in ShCG have normal colors for their types. Consequently, the apparently red colors are not intrinsic but mainly due to the K-effect. As expected from the morphology of the galaxies in ShCG, not many blue galaxies are found in them, with few exceptions like #4 in Shk 278.

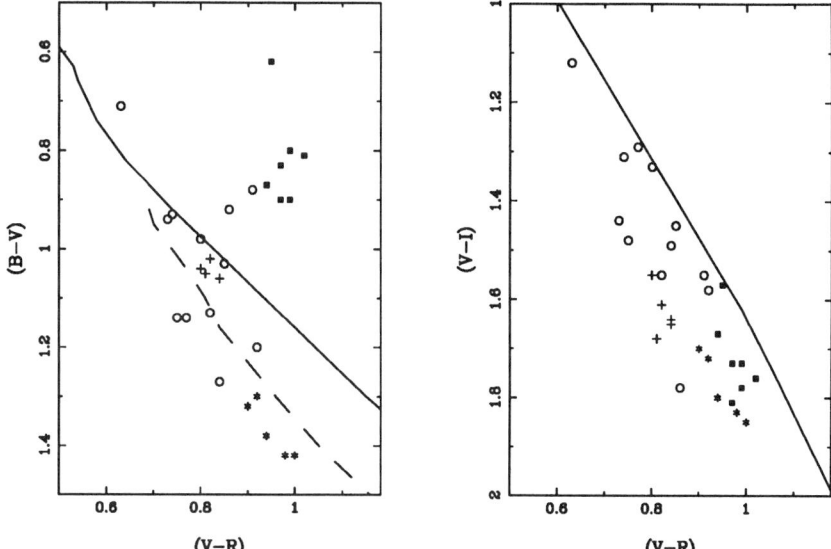

FIGURE 4. Color − color diagrams for member galaxies in Shk groups. Symbols as in Figure 3. Solid line for the main sequence stars and dashed line for giants.

2.3. How luminous are the galaxies?

The comparison with the profiles by Schombert already indicated that the galaxies are rather luminous. The range of the luminosities we have found for the isolated groups shows that essentially all the galaxies are very luminous, with a marked deficiency of faint objects. The results for Shk 81 are again different in the sense of being dominated by a bright galaxy surrounded by significantly fainter objects. As we already pointed out, this is not a truly isolated compact group but the central part of a cluster of galaxies.

Thus the luminosity function of galaxies in ShCG appears to be truncated. We find essentially only high luminosity objects within ShCGs. That deficiency in faint galaxies was also noticed by Kodaira et al. (1990) in other ShCG.

2.4. How metal-rich are the galaxies?

For the galaxies with high S/N spectroscopic information we have measured the Mg_2 metallic indicator. The values are collected in Table 1. The galaxies in the sample have all above-solar metallicities corresponding to old, evolved, bright early type galaxies. This was also found by Amirkhanyan et al. (1988).

2.5. How massive are the galaxies?

The masses of some of the galaxies were estimated from the value of the velocity dispersion calculated from the spectral lines and the photometric parameters, following Poveda (1958). The velocity dispersions were obtained using the cross-correlation technique by Tonry and Davis (1979). In Table 1 we give the values

TABLE 1. Photometric and spectroscopics properties of galaxies in ShCG

Galaxy	σ_V (km s^{-1})	R_E (kpc)	M_{VT} ($M_\odot$)	Mg_2	M/L_B
Shk 362-1	280±40	8.5	1.3×10^{12}	0.30	55
Shk 362-2	172±35	6.5	3.8×10^{11}	0.27	28
Shk 362-3	180±40	5.4	3.3×10^{11}	0.26	27
Shk 362-4	——	9.1	——	0.23	——
Shk 362-5	187±37	3.5	2.3×10^{11}	0.35	30
Shk 38-1	198±20	3.3	2.6×10^{11}	0.26	34
Shk 38-2	152±30	6.9	3.2×10^{11}	——	34
Shk 38-3	138±20	5.3	2.0×10^{11}	0.21	23
Shk 38-4	203±60	3.2	2.7×10^{11}	0.24	51
Shk 81-1	286±60	12	2.0×10^{12}	0.24	80

of the velocity dispersions, the effective radii, as well as the derived masses and the M/L_B ratios. In some cases the superposition of the light distributions of different objects made it difficult to obtain accurate sizes. This is particularly true for the galaxies in Shk 38, where the uncertainties in R_E can reach a factor of two. Note that the values of the M/L_B ratios are again normal for giant ellipticals.

3. THE GLOBAL PROPERTIES OF SHCG

We have estimated the spatial density in ShCG for the groups for which we have redshift information. We find that it ranges from 10^3 to 10^5 galaxies/Mpc^3. This is indeed a high density, comparable only to that found in other compact groups such as HCGs.

TABLE 2. Observed properties of ShCG

Group	N	z	σ_V (km s^{-1})	R_H (kpc)	M_V ($M_\odot$)	T_{CR} (H_0^{-1})
Shk 38	5	0.086	325	14.65	3.4×10^{12}	0.01
Shk 362	5	0.087	164	33.71	2.0×10^{12}	0.04
Shk 278	7	0.120	360	36.33	1.0×10^{13}	0.02

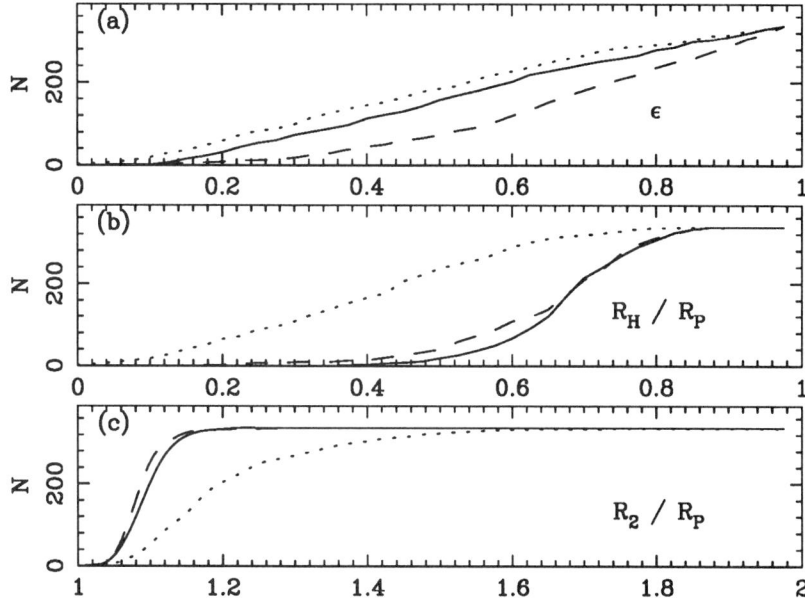

FIGURE 5. Cumulative distribution functions for (a) ellipticity, (b) R_H/R_P and (c) R_D/R_P. Continous line for Shakbazyan groups, dotted line for the N-body simulations and dashed line for the random distributions.

The dynamical characteristics of the 3 groups for which we have enough spectroscopic data are presented in Table 2. The values of the group velocity dispersion, σ_v, are in general not very high, typically similar to the values found for those Hickson groups populated by early type objects. In spite of this, the groups are so small that the crossing time is only a small fraction (about a few percent) of the Hubble time. Thus the study of the ShCG faces the same problems of short crossing times and dynamical scales as do the HCGs.

The masses given in Table 2 are just virial values. A property of the ShCGs, again in common with many of the HCGs, is that the M/L_B ratios are not very high, in the range 30 - 100. This turns out to be not too much higher than for the member galaxies, indicating that the groups could be bound without a significant matter component external to the galaxies. The total mass and M/L_B ratios are similar to those of cD galaxies, which could presumably be the fate of those dense groups.

We have also analyzed the structure parameters of a large fraction (89%) of the whole ShCG sample. We have, however, only considered those parameters which are independent of distance since the redshifts are known for only a small number of groups. These are the ellipticity, and the ratios of the moments (the harmonic radius, R_H, the mean pairwise separation, R_P, and the second order moment) of the relative positions of the galaxies in a group. Then we have compared them with the values found for random distributions and N-body simulations. The results are displayed in Figures 5a, b and c. Let's consider first

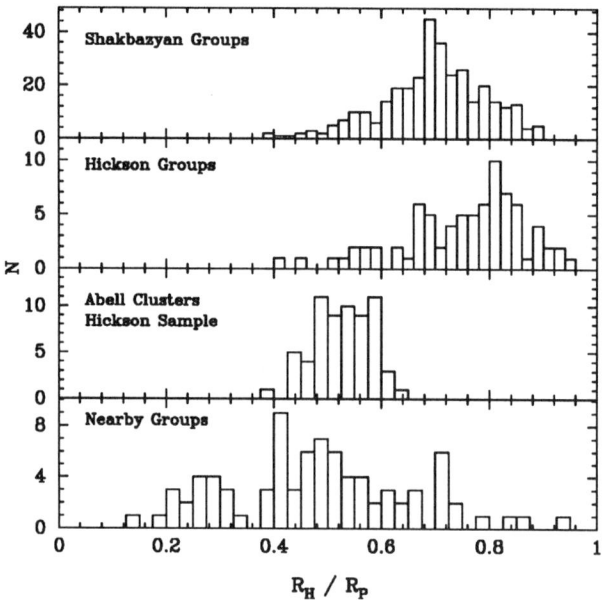

FIGURE 6. R_H/R_P histograms for different samples.

the ellipticity distribution. The actual one for the groups is not very different from what is found from N-body simulations, but is markedly different from random distributions. Note that the real groups show greater ellipticities, a result already pointed out by Vardanyan and Melik-Alaverdyan (1978). As for the moments, the random distribution is much more similar to the real groups. But now the N-body simulations results depart significantly form the real world case.

Finally we present in Figure 6 the R_H/R_P distribution for different samples. Now, the results are similar for ShCGs and HCGs. The tail toward lower values of that parameter in the sample of ShCGs is probably due to the presence of non-isolated groups. This is based on the fact that all the groups in that tail have 7 or more members. These would then be the non-isolated, non-compact associations like as Shk 81, many of them being part of larger aggregates.

4. CONCLUSIONS

The first conclusion refers to the fact that the ShCGs are real entities. This is strongly suggested by the distribution of the structure parameters which shows that they are high density systems, with forms not reproducible by random effects. On the other hand, the departure of those distributions from the predictions of N-points simulations would indicate that tidal effects in the member galaxies are important in setting the dynamical state of the groups.

The dynamical parameters give very short crossing times (a few percent of the Hubble time), which would indicate a very fast dynamical evolution of the groups. Since the galaxies are still there as individuals, it could be said that the observed compact groups are just at the beginning of their lives as compact entities. Unless, the real dynamical processes are not as fast as predicted by most (but not all) the simulations.

The ShCGs contain mainly bright, early-type galaxies. Most of them are ellipticals, with slightly less than one third showing evidence of a faint disk component. Also, in some sense the absence of boxy systems is remarkable. Indeed, the boxiness was sometimes interpreted as a sign of a merger process. If that was the case, it could be concluded that there are no signs of merging processes in the ShCGs. This is, however, not so clear. In fact, other than the lack of strong evidences on the relation between boxiness and merging, the marked deficiency in faint galaxies could be due to the fact that only a few of the members of an originally larger group could cannibalize the other members and survive. The problem is of course that of the dynamical lifetimes of dense configurations. On the other hand, the groups could start already with only bright galaxies, in which case they would reflect the particularities of the galaxy formation in dense regions. The first option seems to be more satisfactory, in particular in explaining the absence of spiral systems. In any case, the problem of the dynamical scale is still there, but perhaps less acute if the number of members of a group is allowed to decrease along the evolution of the aggregate.

The existence of ShCGs makes the problems posed by the compact groups even more acute. Even if only a fraction of the already known compact aggregates are actually physical entities as claimed by some authors (see Mamon in these proceedings), we already have a large number of systems whose existence was not predicted by the models and cannot even be explained.

5. REFERENCES

Ambartsumyan, V.A., Arp, H.C., Hoag, A.A., and Mirzoyan, L.V. 1975, *Astrofizika*, **11**, 193.
Amirkhanyan, A.S., Egikyan, A.G., and Silchenko, O.K. 1988, *Soviet Ast. Lett.*, **14**, 170.
Arp, H.C., Burbidge, G.R., and Jones, T.W. 1973, *PASP*, **85**, 423.
Baier, F.W., Petrosyan, M.B., Tiersch, H., and Shakbazyan, R.K. 1974, *Astrofizika*, **10**, 327.
Baier, F.W., and Tiersch, H. 1975, *Astrofizika*, **11**, 221.
Baier, F.W., and Tiersch, H. 1976a, *Astrofizika*, **12**, 7.
Baier, F.W., and Tiersch, H. 1976b, *Astrofizika*, **12**, 409.
Baier, F.W., and Tiersch, H. 1978, *Astrofizika*, **14**, 279.
Baier, F.W., and Tiersch, H. 1979, *Astrofizika*, **15**, 33.
Börngen, F., and Kalloglyan, A.T. 1974, *Astrofizika*, **10**, 21.
Börngen, F., and Kalloglyan, A.T. 1980, *Astrofizika*, **16**, 599.
Burstein, D., and Heiles, C. 1982, *AJ*, **87**, 1165.
del Olmo, A. 1988, Tesis Doctoral, Universidad de Granada.
del Olmo, A., and Moles, M. 1991, *A&A*, **245**, 27.
Hickson, P. 1982, *ApJ*, **255**, 382.
Kirshner, R.P., and Malumuth, E.M. 1980, *ApJ*, **236**, 366.
Kodaira, K., Iye, M., Okamura, S., and Stockton, A. 1988, *PASJ*, **40**, 533.
Kodaira, K., Doi, M., Ichikawa, S., and Okamura, S. 1990, *P.N.A.O. Japan*, **1**, 283.
Lynds, C.R., Khachikyan, E.E., and Amarkhanyan, A.S. 1990, *Soviet Ast. Lett.*, **16**, 83.
Massey, P. 1977, *PASP*, **89**, 13.

Mirzoyan, L.V., Miller, J.S., and Osterbrock, D.E. 1975, *ApJ*, **196**, 687.
Pence, W. 1976, *ApJ*, **203**, 39.
Petrosyan, M.B. 1974, *Astrofizika*, **10**, 471.
Petrosyan, M.B. 1978, *Astrofizika*, **14**, 631.
Poveda, A. 1958, *Bol.Obs. Tonantzintla and Tacubaya*, **17**.
Robinson L.B., and Wampler, E.J. 1973, *ApJ*, **179**, L35.
Schombert, J.M. 1986, *ApJS*, **60**, 603.
Shakbazyan, R.K. 1973, *Astrofizika*, **9**, 495.
Shakhbazyan, R.K. 1978, *Astrofizika*, **14**, 273.
Shakbazyan, R.K., and Petrosyan, M.B. 1974, *Astrofizika*, **10**, 13.
Tonry, J., and Davis, M. 1979, *AJ*, **84**, 1511.
Vardanyan, R.A., and Melik-Alaverdyan, Y.K. 1978, *Astrofizika*, **14**, 195.
Vennik, J., Kaazik, A., and Amirkhanyan, A. 1982, *Astrofizika*, **18**, 533.

Compact Groups of Galaxies:
The OLF and its Implications

JACK W. SULENTIC AND CARLOS R. RABAÇA
Department of Physics and Astronomy, University of Alabama,
Tuscaloosa, AL 35487, USA

ABSTRACT: The optical luminosity function (OLF) for compact groups is presented using the data in the Hickson (1982) sample. The OLF for late type galaxies in compact groups shows a surprising lack of interaction induced enhancement. A similar lack of starburst activity is seen in the FIRLF based upon the IRAS survey. These results suggest that interaction stimulation of late type galaxies depends more upon the details of the encounter than on the local galaxy density. At the same time, the ellipticals in compact groups are much brighter than expected. The latter result supports the contention that compact groups are physically dense systems. A comparison between the OLF for groups as units and an isolated field elliptical sample reveals no evidence for the expected population of compact group merger remnants. The data suggest that compact groups merge slowly if at all.

1. COMPACT GROUPS: WHAT ARE THEY?

Compact groups of galaxies (hereafter CG's) pose interesting problems for our ideas about galaxy formation and evolution. They represent the densest aggregates of galaxies found outside the cores of rich clusters. These systems of typically 4–10 galaxies (HCG, Hickson 1982) show surface density enhancements over the field (where most HCG's are found) by a factor of from 10^2–10^3 which imply space densities as high as 10^4 pc^{-3}. Best attempts at modeling the groups (e.g., Barnes 1989) lead to some puzzling results and contradictions with observation:

1. They merge quickly and therefore, apparently, cannot be primordial systems. Although an entire group may not coalesce within 1-2 Gyr, partial mergers are expected to begin in one or two crossing times ($t_c \sim 10^8$ years). Since most CG's are systems of 4–5 galaxies, this implies that most will appear as triplets or binaries after a few crossing times (see Rampazzo and Sulentic 1992 for possible evidence). This implies that most observed CG's have recently collapsed out of the loose group environment.

2. Strangely, they have different properties from this environment. The ratio of early to late type systems is much larger in the compact groups (1.0±0.1, Hickson et al. 1989) than in the field (0.22±0.1, van den Bergh 1980).

3. They must also form into a CG before the onset of merging. There is evidence for considerable dynamical stripping and other secular evolutionary effects in groups with no signs of merger activity (Sulentic 1987, Williams and Rood 1987). This suggests that CGs exist for a significant period of time before the onset of merging.

4. Finally their merger remnants should be fairly numerous (e.g., 10^{-4} Mpc^{-3}, Barnes 1989). Little evidence is found for merger remnants or mergers in

progress among CG component galaxies (Zepf et al. 1991 and this paper).

The difficulties with the above scenario have motivated some (e.g., Mamon 1986, 1990) to argue that almost all CG's are chance projections rather than physically dense groups.

We support the view that the majority of CG's are physically dense aggregates. Our evidence (in addition to item 2 above) is at least threefold:

1. CG's are found in regions of generally low galaxy density (Sulentic 1987, Williams and Rood 1987, Hickson et al. 1988). This makes it unlikely that very many could be produced by chance projection.

2. CG's show an underabundance of neutral hydrogen which is difficult to reconcile with chance projection or recent formation hypotheses (Williams and Rood 1987).

3. Finally, numerous CG members show interaction morphology (bridges, tails, luminous halos) ($\sim$ 30%, Hickson 1989), peculiar rotation curves ($\sim$ 50%, Rubin et al. 1991), and tidal truncation in early-type galaxies (Zepf et al. 1991).

If we assume that the CG's are physically dense systems then we can ask whether the most complete sample available to us (Hickson 1982) is representative of CG's. The alternative is that the Hickson catalog represents only the "tip of the iceberg" to some more general population. CG's are either an extremum in the properties of loose groups or they are a distinct class of galaxy aggregate. If the former is true then some selection effect is preventing us from finding the richest CG's (the link between CGs and ordinary groups), since loose groups usually have more members. It has been argued that CG's are transient dense cores in loose groups (Rose 1979) but, again, the bulk of the above evidence rules out this kind of explanation. We favor the interpretation that CG's represent a unique class of clustering. A recent attempt to objectify the Hickson (1982) selection criteria (Prandoni et al. 1991) allows one to represent the selection domain of CG's in a plot of brightest member magnitude versus group diameter. In such a plot the Hickson CG's show no obvious skewing toward one of the boundaries of the domain which might be expected if they formed the tail of a much larger population (with less isolation, or compactness). This skewing might also be expected if the selection criteria were biased towards only some part of a much larger group population. Instead they cluster near the center of the selection domain. Subsequent deep CCD imaging of the Hickson CG's (Hickson et al. 1989) also supports the uniqueness interpretation. The imaging has failed to uncover large numbers of additional group members that might have altered the initial defining characteristics of the groups. New group members were added but not enough to change our description of CG's.

CG's interpreted as dense groups are the ideal place to look for the most extreme forms of interaction induced phenomena. Galaxies in such groups would presumably undergo more frequent close encounters than members of pairs. Their susceptibility to merger would also lead us to expect starburst or nonthermal activity related to mergers in progress. We report here on an analysis of the optical and FIR luminosity functions designed to search for the expected enhancement and for the expected merger remnant population in the field.

2. LUMINOSITY FUNCTIONS

The optical luminosity function is an effective way to compare the level of interaction induced enhancement in galaxy samples. There are two possible approaches to calculating the OLF for CG's. Method 1 involves Monte Carlo simulations of the existing HCG sample (see Mendes de Oliveira and Hickson 1991). Method 2 involves a direct calculation of the OLF and comparison with samples selected in a reasonably similar way (our approach).

The advantage of method 1 is that it allows one to compensate for the small sample size, incompleteness and possible selection biases present in the HCG. One can generate a very large sample of artificial groups that are believed to have the same statistical properties as real CGs. The disadvantages are that it requires assumptions: a) about the parent population of CG galaxies and b) that component magnitudes are uncorrelated. The latter assumption seems dangerous since a correlation is observed between the component magnitudes and colors in binary galaxies (Karachentsev 1982, Dëmin et al. 1984). Method 2 makes no assumption about parent populations or degree of correlation between component magnitudes. It suffers from the limitations and biases of the real sample but attempts to compensate by comparing results with similarly selected samples. Sample incompleteness is not necessarily a problem for a luminosity function calculation. It is the presence of luminosity dependent selection biases that represent a hazard. We adopted the CPG (Karachentsev 1972) and CIG (Karachentseva 1973) as our paired and single galaxy control samples. Our comparisons are concentrated at the bright end of the OLF where the samples are most complete and biases less likely to be serious. All three of our samples were selected visually from the charts of the Palomar Sky Survey. Correction factors were derived to match apparent magnitudes into the B_T system (Hickson et al. 1989). All samples were selected with an isolation criterion which means that the singles, binaries and CG's occupy regions of similar galaxy surface density.

We adopted the classical V_m estimator for our OLF calculation (see Xu and Sulentic 1991 for details). Table 1 lists the relevant parameters for our three samples. Values in parenthesis refer to pairs or groups as units. Sky coverage Ω is rather similar but the degree of completeness (and resultant correction factors ξ) are quite different. The CG sample is the most incomplete with the flattest redshift distribution. We were forced to adopt a fainter magnitude cutoff m_{lim} for the HCG in order to maximize our sample size. The cutoff for the CG sample was based upon the V/V_m test for the CG's as units (combined group magnitudes). Once the magnitude cutoff was adopted we included all known redshift accordant members of this sample in our OLF calculation. This is important because a cutoff based upon group component magnitudes would have biased us towards the higher ranked members of the groups.

Figure 1 shows the resultant OLFs for the galaxy components of the CIG, CPG and HCG samples. The best fit Schechter (1976) functions are superimposed on the data. The relevant parameters in the Schechter fits are summarized in Table 2. We are only concerned here with the α and M_* parameters which quantify the slope of the faint end and the position of the exponential dropoff at the bright end, respectively. The latter parameter most effectively measures any interaction induced luminosity enhancement in the samples. The last two

TABLE 1. Sample Parameters

Sample	m_{lim}	ξ	$\langle V/V_m \rangle$	Number	Ω
HCG as units	18.0 (15.0)	7.1	0.20	307 (68)	5.7
CPG as units	16.0 (14.2)	2.7	0.36	865 (433)	3.852
CIG galaxies	15.0	4.3	0.37	375	3.852

columns of Table 2 give a measure of the goodness of fit of the Schechter function. Figure 1 and Table 2 show that the interacting binary sample (CPG) turns over at a brighter absolute magnitude than the CIG. This indicates an excess of luminous galaxies that is usually ascribed to interaction induced starburst activity. The surprising result is that the HCG sample shows little or no evidence for the excess that is clearly seen for the CPG and confirmed in many independent samples of pairs (see refs. in Xu and Sulentic 1991).

TABLE 2. Schechter Fits to the Total OLF

Sample	α	M_*	χ^2	Significance
CIG ...	-1.4 ± 0.1	-20.17 ± 0.10	13.78	0.50
CPG ...	-1.1 ± 0.1	-20.48 ± 0.06	25.39	0.06
HCG ...	-1.1 ± 0.1	-20.23 ± 0.11	25.51	0.01

It is possible to increase the enhancement observed in the HCG sample as measured by M_*. This can be accomplished by fitting a Schechter function only to the galaxies brighter than -18. It is difficult to justify this selective fit because the fainter galaxies have redshifts consistent with CG membership. The exclusion of the faint galaxies increases the slope of α while at the same time brightening the value of M_* by ≈ 0.3. There are two possible reasons for this: 1) a Schechter fit with the dwarfs included suffers from incompleteness at the faint end and 2) the Schechter function does not fit an enhanced LF as well as an unenhanced one. The latter suggestion is supported by the poor fit of the Schechter function for the CPG compared to the CIG. A good fit to the bright end produces a poor fit to the faint end and vice versa. This is true even for a reasonably complete sample like the CPG. The correlation between the two Schechter parameters is apparently different for enhanced and unenhanced samples. If one rejects the data fainter than -18, the OLF suggests an enhancement at a level similar to (but not greater than) the CPG. This fit also implies a much steeper faint end slope than observed for the CPG and more similar to the CIG (implying a deficit of dwarf members in the HCG). The CIG parameters are similar to those found by Davis and Huchra (1982). The steep

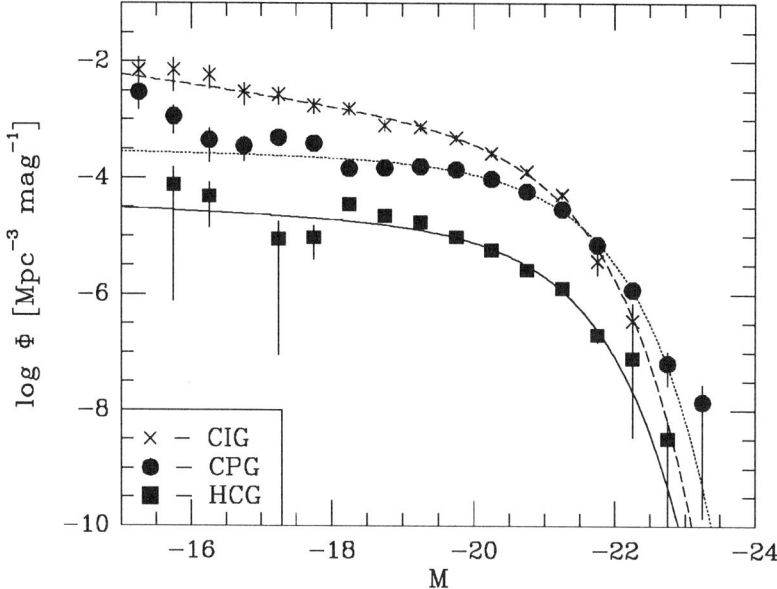

FIGURE 1. OLF for the CIG single (×), CPG pair (•), and HCG compact group (■) samples. Best fit Schechter functions are superposed on the data.

faint end slope is usually attributed to contamination from the local supercluster (Binggeli et al. 1988).

Our CG galaxy sample is large enough to permit a first look at the type specific OLF's. We concentrate here on the E and S subsamples. We find evidence for a significant enhancement of the E components in compact groups. It contributes essentially all of the HCG enhancement observed in Figure 1. A similar, but smaller, effect was previously noted in binaries (Xu and Sulentic 1991). The M_* parameter for HCG ellipticals is a full magnitude brighter than for the CIG sample and 0.8 magnitude brighter than the CPG. Part of this effect could be due to a bias — many of the most distant CG's contain overluminous ellipticals (Mendes de Oliveira and Hickson 1991). However the enhancement remains virtually unchanged when the eight most distant CGs are excluded from the calculation. There is a significant population of elliptical galaxies in compact groups that are as bright as the most luminous cluster ellipticals. On the contrary, surprisingly little enhancement is observed for the spiral components of compact groups. The spiral population in binaries shows a clear optical enhancement (0.3 in M_*). The average spiral component of a binary is a factor of two (0.7 magnitude) brighter than a similar galaxy that is isolated (Karachentsev 1987, Xu and Sulentic 1991). Note that the full effect of the enhancement does not show up in the M_* parameter which is correlated with α and calculated for the complete SS binary sample without correction for accordant redshift nonphysical pairs.

It is tempting to ascribe the observed weakness of an enhancement in CG spirals to some sample bias. The HCG redshift bias should produce the opposite effect unless all high redshift CGs are rich in early type galaxies. Independent evidence for a lack of starburst activity can be found by examining the FIRLF for late type galaxies in the HCG. An earlier attempt to measure the FIRLF argued for the expected starburst enhancement. This analysis did not allow for the resolution limitations of the IRAS database. Many 60 μm and most 100 μm fluxes for compact groups suffer from optical confusion. Sulentic and de Mello Rabaça (1993) recomputed the fractional infrared to optical LF for HCG galaxies and for a control sample. Correction for optical confusion at 60 and, especially, 100 μm yield a result consistent with little or no FIR enhancement in CG spirals. The result is not conclusive since the data allow only an educated guess at the distribution of FIR flux between members of most CG's. It is clear that the modest effect found in the earlier analysis is an upper limit to any possible enhancement.

3. SEARCHING FOR MERGER REMNANTS

A final calculation involves a comparison of the CIG OLF with the LF for CG's as units. If the typical merger of a CG results in an E like galaxy with absolute magnitude similar to the sum of the components then we should see considerable overlap in our comparison. According to the estimate of Barnes (1989) we expect several hundred remnant in the volume of space sampled by the CIG. The overlap is further expected because the CIG represents a reasonable approximation of the single galaxy environments where CG's are found. We find that the OLF for CG's as units is displaced two magnitudes brighter relative to the CIG. No elliptical galaxies brighter than -22 were observed in the CIG while a large fraction of HCG's show this combined luminosity. Figure 2 shows a comparison of the absolute magnitude distribution for these two samples. The CIG should be least biased against this luminous population. Either CG mergers fade considerably ($\geq$ factor of five) or the expected population of CG merger remnants is not observed.

4. CONCLUSIONS AND IMPLICATIONS

Binary galaxies show a stronger enhancement than the components of CG's even though the excess of high redshift CG's would predict a bias in the opposite direction. While the Schechter function does not give a particularly good fit to enhanced OLF's, it quantifies the surprising lack of enhanced spirals and excess of enhanced ellipticals in CGs. The FIRLF provides possible confirmation of the low level of starburst activity in late type members of CGs. We find no evidence for a population of CG merger remnants in our field sample.

The OLFs for CG's and the field are different suggesting that compact groups are not optical projections. The weak enhancement observed for late type galaxies in CG's suggests that interaction stimulation depends more upon the details of the encounter than upon the local galaxy density. If the above is untrue, then some unknown process mitigates the level of optical, FIR, and probably radio, enhancement in late type members of compact groups. The nature and origin of the early type enhancement is in doubt given the lack of evidence that these galaxies are merger remnants (see Zepf *et al.* 1991). Finally

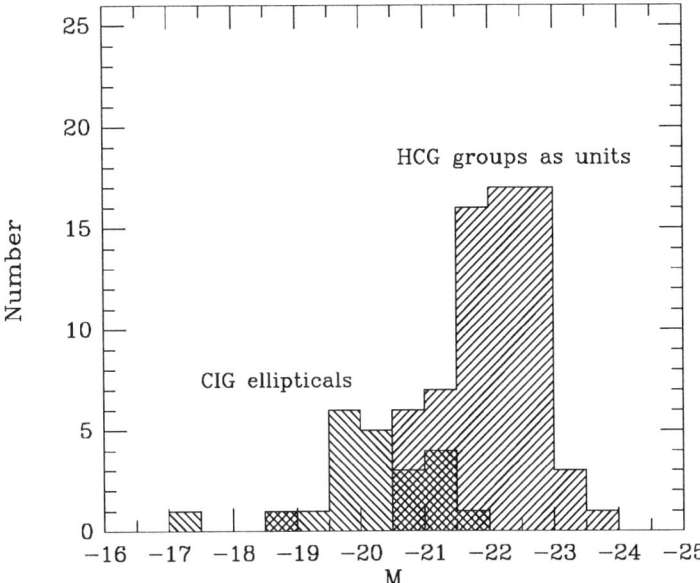

FIGURE 2. Absolute magnitude distributions for compact group combined magnitudes and isolated elliptical galaxies in the CIG.

the lack of a *bona fide* merger population in the groups as well as a population of remnants in the field suggest that CG's merge very slowly if at all.

5. REFERENCES

Barnes, J.E. 1989, *Nature*, **338**, 123.
Binggeli, B., Sandage, A., and Tammann, G. 1988, *ARA&A*, **26**, 509.
Davis, M., and Huchra, J.P. 1982, *ApJ*, **254**, 437.
Dëmin, V., Zasov, A., Dibaǐ, É., and Tomov, A. 1984, *Soviet Ast.*, **28**, 367.
Hickson, P. 1982, *ApJ*, **255**, 382 [HCG].
Hickson, P., Kindl, E., and Auman, J.R. 1989, *ApJS*, **70**, 687.
Hickson, P., Kindl, E., and Huchra, J.P. 1988, *ApJ*, **329**, L65.
Karachentsev, I.D. 1972, *Izv.Spets.Astrofiz.Obs.*, **7**, 1 [CPG].
Karachentsev, I.D. 1982, *Astrophysics*, **17**, 135.
Karachentsev, I.D. 1987, *Double Galaxies*, (Nauka: Moscow).
Karachentseva, V.E. 1973, *Soobshch.Spets.Astrofiz.Obs.*, **8**, 1 [CIG].
Mamon, G.A. 1986, *ApJ*, **307**, 426.
Mamon, G.A. 1990, in *Paired and Interacting Galaxies*, eds. J.W. Sulentic, W.C. Keel, and C.M. Telesco, NASA CP-3098, p. 609.
Mendes de Oliveira, C., and Hickson, P. 1991, *ApJ*, **380**, 30.
Prandoni, I., Iovino, A., Bhatia, R., MacGillivray, H.T., Hickson, P., Palumbo, G. 1992, in *Digitized Optical Sky Surveys*, eds. H.T. MacGillivray and E.B. Thompson, Kluwer, Academic Publishers, p. 361.
Rampazzo, R., and Sulentic, J.W. 1992, *A&A*, **259**, 43.
Rood, H.W., and Williams, B.A. 1989, *ApJ*, **339**, 772.
Rose, J.A. 1979, *ApJ*, **231**, 10.

Rubin, V.C., Hunter, D.A., and Ford, W.K. 1991, *ApJS*, **76**, 153.
Schechter, P.L. 1976, *ApJ*, **203**, 297.
Sulentic, J.W. 1987, *ApJ*, **322**, 605.
Sulentic, J.W., and de Mello Rabaça, D.F. 1993, *ApJ*, **410**, 520.
van den Bergh, S. 1980, *Phil. Trans. Roy. Soc. London*, **296**, 319.
Williams, B., and Rood, H.J. 1987, *ApJS*, **63**, 265.
Xu, C., and Sulentic, J.W. 1991, *ApJ*, **374**, 407.
Zepf, S.E., Whitmore, B.C., and Levison, H.F. 1991, *ApJ*, **383**, 524.

GALAXIES IN COMPACT GROUPS

STEPHEN E. ZEPF

Physics Department, University of Durham, South Road, Durham DH1 3LE, England

ABSTRACT: I review observations of galaxies in compact groups and present evidence that galaxies within the groups are affected by the interactions and mergers which are expected to be common in these regions. The evidence is strongest for properties affected by dynamical perturbations and weakest for the creation of new ellipticals through the mergers of spirals. Taken as a whole, the observations indicate that there are fewer interactions and mergers than predicted by calculations based on the short crossing time observed for the groups. This discrepancy is best explained by a combination of selection effects, initial distribution of orbits, and the distribution of dark matter.

1. INTRODUCTION

Compact groups of galaxies first attracted attention because of their striking appearance (they still rank as some of the most photogenic of astronomical objects). They have also long attracted the interest of dynamicists because of their apparently short time scales for dynamical evolution (e.g., Peebles 1971). Recent interest in compact groups has been driven by the recognition that their high spatial densities and low velocity dispersions imply that mergers and interactions are more common in these groups than other environments at the current epoch. This interest has been encouraged by the systematic survey of Hickson (1982) and subsequent spectroscopic work (Hickson, Kindl, and Huchra 1989) which gives the subject a sound basis from which to begin more detailed studies. Therefore, the compact groups classified by Hickson (hereafter HCGs) appear to be the best available laboratories in which to study how interactions and mergers affect the basic properties of galaxies.

2. EVIDENCE FOR INTERACTIONS AND MERGERS IN COMPACT GROUPS

2.1. Colors of Elliptical Galaxies

The proposal that elliptical galaxies are made through the mergers of spiral galaxies has remained one of the most controversial aspects of the debate on the role of mergers in galaxy evolution. One argument against this proposal is the tightness of the color-magnitude relationship for elliptical galaxies. A tight relationship is not expected if a significant number of ellipticals have recently been formed through mergers because mergers of stellar disks coalesce into the morphology of an elliptical more quickly (≤ 1 Gyr; e.g., Barnes 1990) than the younger, blue stellar populations of spiral galaxies evolve to the older, red stellar populations typical of elliptical galaxies (≥ 2 Gyr; e.g., Charlot and Bruzual 1991).

Thus the merger hypothesis clearly predicts the existence of a population of galaxies with the morphology of an elliptical galaxy but with an unusually blue color. Although a few such galaxies have been observed (e.g., Schweizer et al. 1990), they are rare. For example, only 9 of the 378 elliptical galaxies in the Burstein et al. (1987) sample have $(B-V)_0$ colors more than 0.10 magnitudes bluer than expected from the color-magnitude relationship. However, the current epoch is generally not expected to be the place to look for recent mergers, as most models predict they were much more frequent in the past (e.g., Toomre 1977). Therefore, it is difficult to test the merger hypothesis by studying ellipticals at the current epoch, because most of the putative mergers occur at high redshift.

Compact groups offer a way around this problem because they are the one environment at the current epoch in which mergers are expected to be common. Therefore, we (Zepf, Whitmore, and Levison 1991) undertook a multicolor survey of early-type galaxies in HCGs. We observed 55 elliptical galaxies and 21 S0 galaxies and found that three of the elliptical galaxies and one of the S0 galaxies have colors significantly bluer than expected for their absolute magnitude (i.e. more than 0.10 magnitudes bluer in $(B-V)_0$). A difference of 0.10 magnitudes is more than three times the 1σ uncertainty of our $(B-V)_0$ color as determined both from internal estimates and from comparison to observations of galaxies in the Burstein et al. catalog which we interleaved in our observing program.

In addition to having blue $(B-V)_0$ colors, these four galaxies are also bluer in other optical colors. Moreover, in the two cases for which we have suitable spectra, we observe enhanced Balmer absorption lines indicative of young stars. Another notable feature of these galaxies is that they are bluer in the center than at large radius, which is opposite to the color gradients of most other elliptical galaxies which become redder in the center (e.g., Peletier et al. 1990). Finally, recent aperture photometry by Moles et al. (1992) confirms our small errors in $(B-V)_0$. Thus, the conclusion that the stellar populations of these galaxies are younger than those of most ellipticals seems inescapable.

In addition to finding several unusually blue elliptical galaxies, it also appears that there may be several unusually *red* elliptical galaxies in the HCGs. This was first noted by Zepf et al. (1991), and the red color of at least one of these ellipticals is confirmed by Moles et al. (1992). This result was not anticipated by simple models, and might be explained by dust absorption. However, there is no other evidence for dust in the one confirmed case, which is most notable for being round and having a large isophotal twist.

The primary result of our survey of HCG elliptical galaxies is that 5% have unusually blue colors. In order to interpret these results, we developed models which combine the results of N-body simulations with simple models of the evolution of stellar populations in order to predict how many unusually blue ellipticals are expected in the "standard picture" of the evolution of compact groups (Zepf and Whitmore 1991). These models indicate that about 15% rather than the observed 5% of the ellipticals in compact groups should have unusually blue colors as a result of galaxy mergers. This discrepancy between the models and the observations suggests that either 1) former galaxy mergers do not generally resemble elliptical galaxies but rather some other type of galaxy, or 2) the time scale for the dynamical evolution of compact groups through merging has been underestimated.

2.2. Ongoing Mergers

Another critical observational constraint on models of the evolution of galaxies in compact groups is the fraction of galaxies which are currently merging. This observation is complementary to that of the fraction of elliptical galaxies which are blue because it is the ongoing mergers which will presumably make unusually blue ellipticals in the future. Unfortunately, the identification of an ongoing merger is operationally more uncertain than the identification of an elliptical galaxy with blue colors. However, the importance of the answer to understanding the evolution of galaxies in compact groups is great enough that it is worthwhile to make the best possible determination of the number of ongoing mergers in HCGs.

There are a number of possible ways to identify merging systems. In a recent paper (Zepf 1992), I utilized three independent methods to determine the number of ongoing mergers in HCGs. The first method identifies mergers by the extended tidal tails produced during close encounters of spiral galaxies (e.g., Toomre and Toomre 1972). Because encounters of galaxies in HCGs will nearly always be at low velocities, these tails are very likely to indicate a merging system. Therefore, images of the HCG galaxies were examined, and those systems with clear tidal tails (like those identified by Toomre 1977) were classified as ongoing mergers. Approximately 19 of the 320 galaxies surveyed were classified as mergers by this method. After correcting for the bias against mergers of early-type galaxies inherent in this method, I find that roughly 7% of the galaxies in HCGs have morphologies indicative of a merging system.

The second technique is based on the severe dynamical disturbance produced by a merger. Specifically, field objects commonly identified as merging systems (e.g., NGC 7252) appear to have "sinusoidal" rotation curves (i.e. they go through zero three times) whereas this feature is not found in the rotation curves of spiral galaxies in other samples (Rubin *et al.* 1985, Rubin, Whitmore, and Ford 1988). Therefore, of the 33 HCG late-type galaxies whose dynamics were studied by Rubin, Hunter, and Ford (1991), the two objects with clearly sinusoidal rotation curves are identified as mergers, with a third object possibly belonging in this category.

The final method for identifying mergers uses the IRAS far-infrared data for galaxies in compact groups (Hickson *et al.* 1989) and relies on the observation that merging galaxies have warmer far-infrared colors than non-merging galaxies (e.g., Mazzarella, Bothun, and Boroson 1991). The far-infrared color is used in preference to the far-infrared to optical luminosity ratio because Mazzarella *et al.* and others have shown that it is a more robust indicator of merging. The fraction of HCG galaxies which are merging is determined by comparing the distribution of the far-infrared color of HCG galaxies to samples of galaxies believed to be merging (e.g., the Toomre 1977 sample) as well as to large samples of normal galaxies (e.g., the study of the IRAS data for galaxies in the UGC catalog by Bothun, Lonsdale, and Rice 1989) This comparison indicates that about 5.5% of the HCG galaxies are currently merging, with lower and upper limits at the 90% confidence level of 2% and 9% respectively.

Each of these independent methods indicates that roughly 7% of the galaxies in HCGs are currently merging. The agreement among the various methods is probably somewhat fortuitous, since all of the techniques are subject to significant uncertainties, both statistical and systematic. However, the methods tend to agree with one another in the classification of specific galaxies which lends weight to the results and suggests that the percentage of galaxies in HCGs which are merging is close to 7%. Further details of this work can be found in Zepf (1992).

Although the level of merging in compact groups is significantly higher than the level found in field samples using the same techniques, the fraction of mergers is still relatively modest. Therefore, the conclusion based on both the fraction of galaxies which are currently merging and the fraction of ellipticals with unusually blue color is that the correct model for compact groups is one in which the dynamical time scale of evolution through merging is greater than that based on the observed crossing time. Moreover, a comparison of the expected color of the final product of compact group evolution to the observed color distribution of elliptical galaxies also favors a long time scale for the evolution of HCGs through mergers. These arguments and additional constraints are discussed in more detail in Zepf and Whitmore (1991).

2.3. Structure and Dynamics of Elliptical Galaxies

Although dramatic mergers which result in a transformation of galaxy type appear to be relatively rare in HCGs, many less dramatic interactions and accretion events may affect the galaxies within the groups. In elliptical galaxies, such events can manifest themselves as features in the isophotal structure and the stellar kinematics of the galaxies. Therefore, we undertook a thorough study of the structure and dynamics of the elliptical galaxies in compact groups (Zepf and Whitmore 1992).

Fortunately, there is a large body of recent work classifying the isophotal shapes of elliptical galaxies in other environments (e.g., Bender *et al.* 1989, Peletier *et al.* 1990). These studies find that the isophotes of most elliptical galaxies deviate slightly from perfect ellipses with only about one-third of the galaxies having isophotes which are perfectly elliptical to within a fraction of a percent. The most common deviations are a small excess of light along the major axis (a "disky" elliptical) or an excess of light at 45° to the major axis (a "boxy" elliptical), with a small number of ellipticals having irregular isophotes which do not fit into any of the above categories.

Classifying the isophotal shapes of 32 HCG elliptical galaxies in the same way as these other studies, I find that the elliptical galaxies in compact groups have significantly different isophotal shapes than elliptical galaxies in other environments (Zepf 1991). Specifically, galaxies with small deviations from elliptical isophotes are less common in HCGs, ellipticals with irregular isophotes are more common in HCGs, and ellipticals with boxy isophotes are less common in HCGs (see Table 1). The observation that ellipticals in compact groups often have irregular isophotes is suggestive of more frequent interactions in HCGs because these deviations are often associated with tidal distortions or dust lanes. However, the lack of ellipticals with boxy isophotes is surprising in this context because boxy isophotes have been suggested as a sign of a past merger (e.g., Bender *et al.* 1989). This lack of boxy ellipticals is unlikely to be due to classification errors, since Bettoni (this volume) has independently reached the same conclusion. The most likely explanation for this observation is that the relationship between boxiness and merging is weak, as suggested by Barnes (1992).

TABLE 1. Distribution of Isophotal Shapes of Elliptical Galaxies

Sample	No.	Boxy	Disky	Irregular	Symmetric
Bender et al.	47	17%	36%	17%	30%
Peletier et al.	39	26%	21%	21%	33%
Zepf (HCGs)	32	6%	34%	50%	9%

The stellar kinematics might also be affected by the environment. In order to look for such changes, we obtained central velocity dispersions of 20 ellipticals in compact groups (Zepf and Whitmore 1992). We find that ellipticals in HCGs have lower velocity dispersions than ellipticals in other environments with the same effective radii, surface brightnesses, magnitudes, and colors. Therefore, the elliptical galaxies in compact groups lie off of the fundamental plane defined by other elliptical galaxies.

The sample size is unfortunately small, and a firm conclusion must await additional data which we are now obtaining. However, there are several factors which support the hypothesis that elliptical galaxies in HCGs have low velocity dispersions. The first is that distance errors are unlikely to be a factor since the HCG ellipticals have low velocity dispersions for their colors. The second factor is that the central velocity dispersions correlate with the isophotal shape, such that disky ellipticals have lower velocity dispersions than boxy ellipticals. This trend not only supports the reality of the low velocity dispersions, but also suggests that the same physical phenomenon may be responsible for the differences in isophotal shape and the low velocity dispersions.

Therefore although dramatic mergers which transform morphological type appear to be rare, some tidal effects are occurring which have important implications for the structure and dynamics of the galaxies in compact groups. It is interesting to note that similar effects to those described here for elliptical galaxies also appear to be observed for spiral galaxies in compact groups (Rubin et al. 1991). Specifically, although only a few of the spirals have the sinusoidal rotation curves indicative of a merger, the majority have some type of less disruptive dynamical peculiarity. In addition, HCG spirals may also have a lower maximum amplitude of their rotation curves than equivalent field galaxies.

3. THE DYNAMICS OF COMPACT GROUPS

Since observations indicate that compact groups evolve through merging more slowly than predicted by simple models, it is interesting to consider ways in which the model predictions can be modified. Perhaps the most obvious place to start is with the observational basis for the dynamical time scale. The median apparent crossing time for the HCGs is 0.016 H_0^{-1} (Hickson et al. 1992). It is this value which gives a time scale for merging that is shorter than required by the observations, even after accounting for the fact that the merging time scale is much longer than the crossing time (White 1990).

The most drastic way to lengthen the evolutionary time scale of HCGs is to accept Mamon's (1986) proposition that most of the compact groups are projections within loose groups and clusters. Although projection among completely unrelated galaxies is ruled out by the enhanced frequency of interactions reported in previous sections, Mamon (1987, 1990) later suggested that many of the projections will be around true pairs or triplets. This latter hypothesis is much more robust because it naturally explains the evidence presented in §2.2 that currently merging systems are more common in compact groups than in other environments. However, it seems more difficult for this hypothesis to accommodate the observations described in §2.3 which indicate that dynamical perturbations are common in many galaxies in HCGs and suggest that most galaxies know they are in a compact groups. The various correlations between the properties of individual galaxies and the groups as a whole also provide circumstantial evidence in favor of the idea that the observed properties of the HCGs accurately reflect their true properties (Hickson and Rood 1988).

The most persuasive way to show that compact groups are real physical entities is to detect an intergalactic medium encompassing the entire group. This intragroup material might take the form of cold gas, hot gas or a diffuse distribution of stars. Unfortunately, such observations are not yet common. As discussed by Williams (these proceedings) at least three of the seven groups mapped in HI have a cloud which encompasses the entire group. Similarly, of the five groups which were observed by Einstein, two appear to have emission attributable to a diffuse intragroup medium (Bahcall, Harris, and Rood 1984). The promise of the ROSAT satellite to substantially improve our knowledge of the distribution of hot gas around HCGs appears to be borne out by initial results from a study of one group which shows substantial extended emission (Ponman, private communication). The only quantitative search for intragroup starlight yet undertaken is that of Rose (1979) in which no intergalactic light was found on deep photographic plates of two groups. Clearly, a more comprehensive survey using large format CCDs is called for (see discussion in the following section on future research).

Although the numbers are small, the direct detections of an intragroup medium imply that many groups have such halos because any detections are difficult to achieve. Thus, the evidence suggests that most compact groups are real physical entities in the sense that there is a strong link between the observed properties of the groups and their true physical properties. However, these arguments against the hypothesis that HCGs are merely loose groups seen in projection do not imply that observed properties of the groups are unaffected by selection biases. In fact, the selection of compact groups naturally leads to the identification of groups whose observed crossing times are an underestimate of their true crossing times. This bias must exist at some level, although ascertaining the level of the effect requires a detailed knowledge of the multiplicity and spatial distribution of galaxies within groups and is very sensitive to the assumptions made about these quantities.

Two other effects in addition to selection biases are likely to play a major role in determining the time scale of the dynamical evolution of compact groups.

The first of these is the distribution of dark matter. In general, models of the dynamical evolution of compact groups have made the simplest assumption about the dark matter within compact groups, which is that it is distributed around individual galaxies. However, if the dark matter is distributed throughout the group rather than attached to individual galaxies, the cross-section of the galaxies is reduced, and the dynamical evolution of the group generally takes longer (e.g., Mamon 1987). Although the stability of the galaxies themselves limits how much dark matter can be removed from individual halos, considerable room may still exist to retard merging by smoothly blending the individual dark matter halos into a common group halo.

An additional way to alter the predictions of simple models is to change the initial orbits of the galaxies. The sensitivity of the time scale for dynamical evolution to the initial orbits of the galaxies was dramatically demonstrated by the simulation of Governato, Bhatia, Chincarini (1991). They showed that at least one set of initial conditions gives a compact group which lasts a Hubble time.

The sensitivity of the time scale of the dynamical evolution of compact groups to the various effects described in this section indicates that there is a long way to go before we understand the compact groups we observe. The root of the problem is the lack of understanding of the formation of compact groups and the relationship between compact groups and loose groups. Until these processes are better understood, the appropriate initial conditions with which to begin simulations of compact groups will be uncertain and the allowed range of dynamical time scales will remain large.

4. FUTURE RESEARCH

The central unanswered question of research related to compact groups is a very basic one - how did they form? The most promising approach to this problem appears to be placing compact groups in the larger context of cosmological simulations. In this way, the formation of compact groups and their relationship to loose groups can be understood in various cosmological scenarios. Such an effort will also provide appropriate initial conditions for detailed simulations of individual compact groups.

The observational subject which seems to offer the best possibility for a significant advance is the search for intragroup starlight. Although the presence of intergalactic light is clearly predicted by models of the evolution of structures such as compact groups (e.g., Carlberg 1990), the primary search for this feature remains the study by Rose (1979) of two groups using photographic plates. Despite the enormous technical advances for this type of work, no new quantitative studies have been published. In collaboration with Mendes de Oliveira, I hope to play a role in remedying this situation by undertaking a deep imaging search for intragroup light using large format CCDs.

Acknowledgements: Much of my work in this field has been in collaboration with Brad Whitmore. I would also like to acknowledge the financial support of an SERC postdoctoral fellowship.

5. REFERENCES

Bahcall, N.A., Harris, D.E., Rood, H.J. 1984, *ApJ*, **284**, L29.
Barnes, J.E. 1990, in *The Dynamics and Interactions of Galaxies*, ed. R. Wielen, Berlin: Springer, p. 186.
Barnes, J.E. 1992, *ApJ*, **393**, 484.
Bender, R., Surma, P., Dobereiner, S., Möllenhof, C., and Madejsky, R. 1989, *A&A*, **217**, 35.
Bothun, G., Lonsdale, C., and Rice, W. 1989, *ApJ*, **341**, 129.
Burstein, D., Davies, R.L., Dressler, A., Faber, S.M., Stone, R.P.S., Lynden-Bell, D., Terlevich, R.J., and Wegner, G. 1987, *ApJS*, **64**, 601.
Carlberg R.G. 1990, *ApJ*, **350**, 505.
Charlot, S., and Bruzual, G. 1991, *ApJ*, **367**, 126.
Governato, F., Bhatia, R., and Chincarini, G. 1991, *ApJ*, **371**, L15.
Hickson, P. 1982, *ApJ*, **255**, 382.
Hickson, P., and Rood, H.J. 1988, *ApJ*, **331**, L69.
Hickson, P., Kindl, E., and Huchra, J. 1989, *ApJ*, **331**, 64.
Hickson, P., Menon, T.K., Palumbo, G.G.C., and Persic, M. 1989, *ApJ*, **341**, 679.
Hickson, P., Mendes de Oliveira, C., Huchra, J.P., Palumbo, G.C.C. 1992, *ApJ*, **399**, 353.
Mamon, G.A. 1986, *ApJ*, **307**, 426.
Mamon, G.A. 1987, *ApJ*, **321**, 622.
Mamon, G.A. 1990, in *Paired and Interacting Galaxies*, eds. J.W. Sulentic, W.C. Keel, and C.M. Telesco, NASA, CP-3098, p. 619.
Mazzarella, J.M., Bothun, G.D., Boroson, T.A. 1991, *AJ*, **101**, 2034.
Moles, M., del Olmo, A., Perea, J., Masegosa, J., Marquez, I., and Costa, V. 1992, preprint.
Peebles, P.J.E. 1971, *Physical Cosmology*, Princeton University Press.
Peletier, R.F, Davies, R.L., Illingworth, G.D., Davis, L.E., and Cawson, M.C. 1990, *AJ*, **100**, 1091.
Rose, J.A. 1979, *ApJ*, **231**, 10.
Rubin, V.C., Ford, W.K. Jr., Thonnard, N., and Burstein, D. 1985, *ApJ*, **298**, 81.
Rubin, V.C., Whitmore, B.C., and Ford, W.K. Jr. 1988, *ApJ*, **333**, 522.
Rubin, V.C., Hunter, D.A., and Ford, W.K. Jr. 1991, *ApJS*, **76**, 153.
Schweizer, F., Seitzer, P., Faber, S.M., Burstein, D., Dalle Ore, C.M., and Gonzalez, J.J. 1990, *ApJ*, **364**, L33.
Toomre, A. 1977, in *The Evolution of Galaxies and Stellar Populations*, eds. B. Tinsley and R. Larson, New Haven, Yale University Observatory, p. 401.
Toomre, A., and Toomre, J. 1972, *ApJ*, **178**, 623.
White, S.D.M. 1990, in *The Dynamics and Interactions of Galaxies*, ed. R. Wielen, Berlin, Springer, p. 380.
Williams, B. 1992, these proceedings.
Zepf, S.E. 1991, Ph.D. Thesis, Johns Hopkins University.
Zepf, S.E. 1992, *ApJ*, submitted.
Zepf, S.E., and Whitmore, B.C. 1991, *ApJ*, **383**, 542.
Zepf, S.E., Whitmore, B.C., and Levison, H.F. 1991, *ApJ*, **383**, 524.
Zepf, S.E., and Whitmore, B.C. 1992, *ApJ*, in preparation.

Simulations of Compact Groups of Galaxies: Some Preliminary Results

E. ATHANASSOULA
Observatoire de Marseille, 2 Place Le Verrier,
F-13248 Marseille Cédex 4, France

J. MAKINO
Department of Information Sciences and Graphics, College of Arts and Sciences, University of Tokyo, Komaba, Meguro-ku, Tokyo 153, Japan

ABSTRACT: We present preliminary results of N-body simulations of the evolution of compact groups. The groups consist initially of five 'elliptical galaxies' and the final product of the simulation is one merger remnant. We find that groups with a common dark halo evolve much faster than groups with individual halos and a dark matter distribution similar to that of the light. The difference in evolution time between groups with equal mass galaxies and groups with unequal mass ones is not as important.

1. INTRODUCTION

As was made evidently clear during this meeting, the structure and the dynamical evolution of compact groups of galaxies are far from being fully understood. Indeed these objects raise a number of interesting questions, in particular concerning their evolution and the time scale in which the galaxies that constitute them will merge. Progress in our understanding is possible with the help of self-consistent N-body simulations with enough particles per galaxy that the main physical effects, the ones that are more important for the evolution of the group, are adequately described. However only few such studies have been made so far. For example there are only a few simulations to add to the list of Mamon (1990), although it was compiled more than two years ago, while a number of the entries in his list have too few particles per galaxy to be realistic.

For this reason we ran a number of simulations hoping their analysis would help clarify a few of the main issues. In particular we would like to understand the effect of a common background halo, and the effect of the existence of a spectrum of masses on the evolution of the group, and to study violent relaxation and the structure of the merger remnant.

Not many observational constraints exist on the amount of dark matter present in compact groups, nor on its spatial distribution. Concerning the latter there are two alternatives. Dark halos could be concentrated around individual galaxies, which we will hereafter call individual halos, or they could encompass the whole group, in which case we will call them common halos. It should, in principle, be possible to distinguish between the two alternatives with the help of X-ray observations. Until the issue has been clarified, however, one should keep an open mind and consider both alternatives.

As their name suggests it, compact groups are very compact with a mean intergalactic separation of 40 h^{-1} kpc (Hickson 1990). Thus if the galactic halos have an extent of say 2 to 3 times that of the optical parts then they would interpenetrate. Considering unperturbed, spherically symmetric and yet interpenetrating halos is not a very realistic initial condition. One would have in such cases to worry about how the galaxies got to that position and start the simulation from the progenitor of the compact group. That would mean including more uncertainties in an uncertain enough situation. We have thus decided to concentrate initially on only two undeniably extreme cases, which should bracket reality. On the one hand halos encompassing the whole group, hereafter denoted for brevity by CH (common halos), and on the other compact individual halos, whose mass distribution follows that of the light, hereafter denoted for brevity by CIH (compact individual halos). The evolution of these two extremes should bracket the case with more extended halos, which we shall consider elsewhere.

The question of the influence of a common halo has already been addressed by previous studies. The simulations of Navarro et al. (1987) do not, strictly speaking, refer to compact groups, since they model either 50 or 100 galaxies, while their average intergalactic distance ranges between 10 and 30 galactic radii. Furthermore galaxies were modelled as mass points and a 'criterion' was adopted to decide whether they had merged during an encounter. Nevertheless it is interesting to note that in these simulations introducing a common halo amounted to slowing down the merging process. A similar conclusion was reached by Cavaliere et al. (1982). Barnes (1985) also discussed this issue, now considering 5 to 10 galaxies per group, and representing each galaxy by a number of points, varying between 75 and 100 depending on the case. The fact that he agrees with the result of Navarro et al. gives thus much more weight to this conclusion.

The question of the effect of a spectrum of masses for the galaxies was addressed by Governato et al. (1991), who found merging times of the order of a Hubble time, i.e. much longer than what was found by other studies. It is thus of interest to test whether this is due to the fact that they used unequal masses.

The most realistic simulation of a group evolution is undeniably that of Barnes (1989). Unfortunately, since only one simulation was performed, it is not possible to derive from it the influence of the halo distribution and/or of the unequal masses.

Our project is in its starting phase. Even though we have neither run enough simulations to cover, albeit roughly, the relevant parameter space, nor have we fully analysed the existing runs, some discussion is timely.

2. SIMULATIONS

In all our simulations the group consisted initially of five 'elliptical galaxies', modelled by isotropic Plummer spheres. The reason we avoided spirals is twofold: in order to reduce the dimensions of the relevant parameter space (since in the case of discs one should also take into account the relative orientation of the discs during the encounter (Barnes 1992)) and to be able to use less particles per galaxy. In the simulations presented here we adopted a visible-to-total mass ratio of 1/3. The common halo, whenever present, was also modelled with a distribution of particles and was therefore responsive. For the simulations with unequal mass galaxies the mass ratios used were $1 : \sqrt{2} : 2 : 2\sqrt{2} : 4$, so that enough particles would be contained even in the smallest galaxy. The velocity dispersion of all galaxies is the same at the start of the simulation.

We use the Plummer model also to represent the distribution of galaxies in the group. The total mass of the group is 20 for all models. The binding energy of the group, excluding the internal binding energy of the individual galaxies, is 5. Therefore, the velocity dispersion within a group is $\sqrt{2}|E|/M = 1/\sqrt{2}$, where E and M are the binding energy and the total mass of the group. G=1 in our units. The initial positions of the galaxies are always the same and are shown in Fig. 1. The mean intergalactic separation is roughly 36 length units. We have not varied the initial positions in order not to increase needlessly the dimensionality of the relevant parameter space, since it is obvious that configurations with shorter intergalactic distances will merge faster.

TABLE 1. List of runs

Run number	Masses	Run time	E/M	σ_g/σ_c	Merging time	halo	comments
1	E	280	-1/4	1	220	CH	
2	U	500	-1/4	1	280	CH	
3	E	760	-1	2	600	CIH	
4	U	400	-1	2	—	CIH	
5	E	630	-1/4	1	400	CIH	
6	E	400	-1/4	1	?	CH	
7	E	600	-1/4	1	600	CIH	
8	E	300	-1/4	1	180	CH	radial orbits
9	E	300	-1	2	220	CH	
10	E	340	-1/4	1	240	CH	
11	E	160	-1	2	60	CIH	radial orbits
12	E	500	-1/4	1	240	CH	maximum rotation
13	E	220	-1	2	—	CIH	maximum rotation

Table 1 lists all our runs in column 1, and the time during which the simulation was run in column 3. Column 2 specifies whether the galaxies had equal (E), or unequal (U) masses. Columns 4 and 5 show the binding energy per unit mass of the individual galaxies and the ratio of the galaxy-to-group velocity dispersion respectively. For the distribution of particles in the background halo, we used the same Plummer model as for the distribution of the galaxies. The velocity dispersion of the halo is chosen so that the total system is in virial equilibrium. Column 7 contains information on the type of halo used in the simulation, and column 8 some comments on the specific initial conditions. For runs 8 and 11 all galaxies were initially on roughly radial orbits. This we did by changing the direction of the velocity vector of each galaxy so that it pointed to the center of mass of the group, while its magnitude was conserved. For models 12 and 13 the velocities were chosen to give rapid rotation to the group. This we did by changing the velocity vectors so that they were perpendicular to both the z-axis and the position vector and then the angular momentum around the

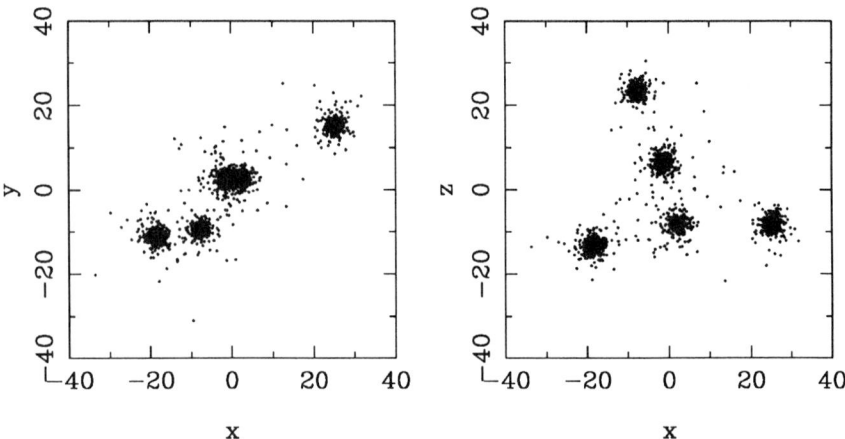

FIGURE 1. Two orthogonal views of the initial conditions for run 1

z-axis became maximum.

We followed the evolution of our groups with a direct N-body code on a dedicated back-end processor, GRAPE-3 (Okumura et al. 1992). This has a peak speed of the order of 10 Gflops so that we could use a sufficiently large number of particles - of the order 20 000 to 30 000 depending on the run - to follow the evolution adequately.

3. RESULTS AND DISCUSSION

Column 6 of Table 1 gives an eye-estimate of the merging time. As such we give the time at which all galaxies in the group have merged into one object, and we have estimated it simply by looking at the evolution of the simulation from two orthogonal points of view. The accuracy of this estimate can obviously not be better than the time interval with which the results were stored (10 or 20 time units depending on the run), but is probably considerably worse, since when exactly the last galaxy has completely merged with the rest is often a matter of opinion. There are of course less subjective ways for making that estimate (e.g. Barnes 1985, Murtagh and Heck 1987) and they will be used in a future analysis, yet for the point we want to make here the accuracy is sufficient. The horizontal lines in the third column for runs 4 and 13 mean that the simulation was not run long enough for merging to occur, while for run 6 the intermediate time steps were inadvertently erased from disc, so that the merging time could not be estimated. It can be seen that the merging time is of the order of a few to a dozen or so dynamical times as defined by Barnes (1985) and White (1990).

One thing is immediately obvious from the entries in Column 6. Namely that the time necessary for the simulations with individual halos to merge is much longer than that necessary for the ones with a common halo. The existence of a common halo can influence the evolution in two ways, antagonistic in the sense of their results. The dynamical friction it provides will slow down the

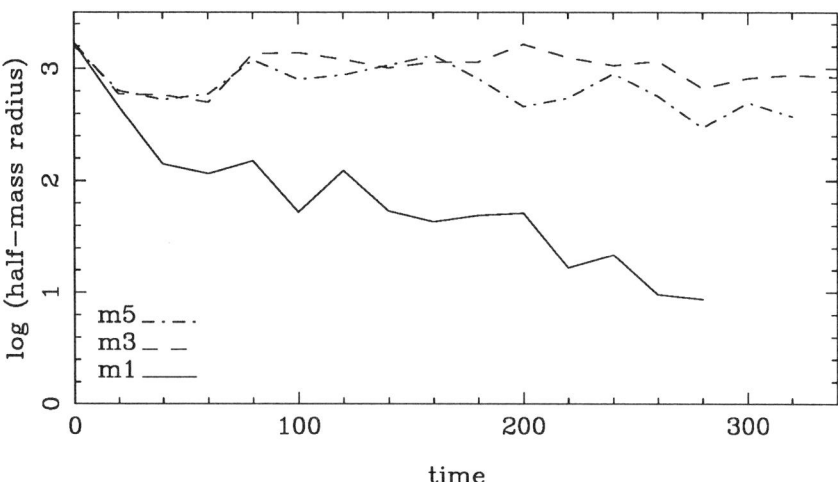

FIGURE 2. Evolution of the half mass radius of the visible matter for runs 1, 3 and 5

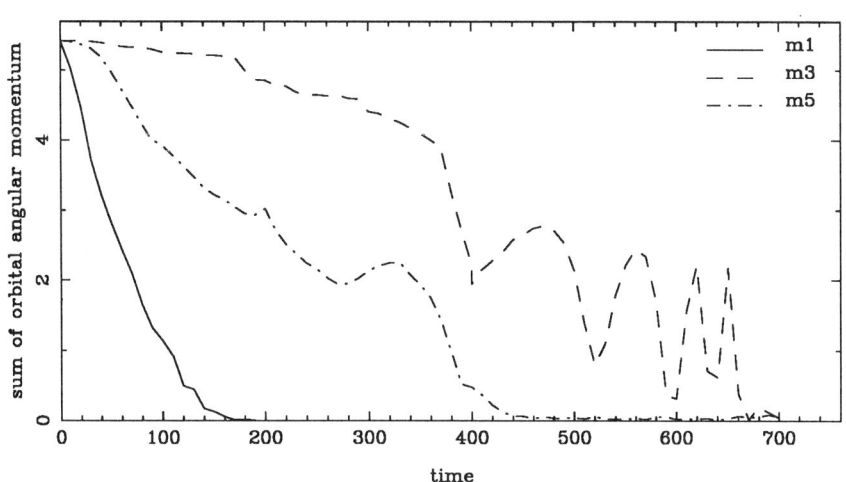

FIGURE 3. Evolution of the total orbital angular momentum for runs 1, 3 and 5

individual galaxies and cause them to spiral in into the center, while the tidal stripping of the outer parts of the galaxies will decrease their merging cross-section and will slow down the evolution. Our simulations indicate that the first effect is predominant. Once the galaxies have spiraled to the center merging will proceed rapidly.

Figure 2 shows the time evolution of the half mass radius of the visible matter for runs 1, 3, and 5, and Figure 3 the evolution of the sum of the galaxies' specific orbital angular momentum for the same runs. For the run with the dark halo distributed over the cluster (run 1) both the orbital angular momentum and the half-mass radius decrease rapidly, the latter roughly exponentially. This is due to the effect of the dynamical friction of the dark halo on the galaxies, which causes them to lose their orbital angular momentum and sink towards the center.

In runs 3 and 5, where the dark matter is associated with individual galaxies, there is no dynamical friction from the halo, and the decrease of the sum of the orbital angular momenta and the shrinking of the half-mass radii are much slower. In these runs the evolution of the group is driven by the inelastic encounters between galaxies. The inelasticity of the encounter is stronger for run 5 than for run 3, since the radius of the individual galaxies in run 5 is larger that that in run 3. For this reason the decreasing of the half-mass radius and of the total orbital angular momentum is faster in run 5 and the merging time is shorter.

It is easy to understand the difference between our results and those of Navarro *et al.* (1987). In their simulations the common halo is modelled as a rigid potential, so they can not include the deceleration and subsequent inwards decay of galaxies due to the dynamical friction of the common halo. Thus they find that the collisions are less 'hard' in the case of common halos, i.e. they have larger relative velocities at the closest approach of galaxies, which in turn leads to less merging according to their 'merging criterion'. The simulations of Cavaliere *et al.* (1982) can also be expected to suffer from the neglect of dynamical friction since the common halo was modelled by 20 equal mass highly softened mass ponts, i.e. of the same order as the number of galaxies. The difference with Barnes' (1985) results can also be understood, since in his simulations the halos have a considerably larger extent than the visual part of the galaxies. Thus, to keep the total mass constant, runs with more mass in the common halo had less mass in the individual halos, and, since the individual halos were considerably more extended than the visual part of the galaxies, that reduced the cross-section of the galaxies (without their velocity dispersion having been changed).

It is obvious that our result depends crucially on the adopted extent of the individual halos, since more extended halos would increase the merging cross section of the galaxies. Because of the short mean intergalactic distances in compact groups, extended individual galactic halos would soon merge into one common halo. Simulations of such groups present a number of difficulties as to their initial conditions and will only be investigated in a future paper.

Although we have an estimate of the merging time for only one run with unequal masses, its value indicates that there is no important difference between the merging times of runs with unequal masses and those with equal ones. The result of Governato *et al.* (1991) hinges on the fact that the orbit of the two massive galaxies in the absence of the small ones was such as to preclude any close encounters for times much longer than a Hubble time, while the perturbation due to the small galaxies was relatively small, since these represented only 25 % in mass. The question now arises whether such configurations are frequent enough to account for the existing compact groups. This, however, can only be answered with the help of cosmological N-body simulations.

Acknowledgements. E. A. gratefully acknowledges the hospitality of the College of Arts and Sciences of the University of Tokyo, where these simulations were run, and a CNRS-JSPS exchange grant which made that trip possible.

4. REFERENCES

Barnes J. E., 1985, *MNRAS*, **215**, 517.
Barnes J. E., 1989, *Nature*, **338**, 132.
Barnes J. E., 1992, *ApJ*, **393**, 484.
Cavaliere A., Santangelo P., Tarquini G., and Vittorio N. 1982, in *Clustering in the Universe*, eds. D. Gerbal and A. Mazure, p. 25, Editions Frontieres, Gif sur Yvette.
Governato F., Bhatia R., and Chincarini G. 1991, *ApJ*, **371**, L18.
Hickson P. 1990, in *Paired and interacting Galaxies*, eds. J.W. Sulentic and W.C. Keel, p. 77, NASA.
Mamon G.A. 1990, in *Paired and interacting Galaxies*, eds. J.W. Sulentic and W.C. Keel, p. 609, NASA.
Murtagh F., and Heck A. 1987, *Multivariate Data Analysis*, Reidel Pub.
Navarro J., Mosconi M.B., and Garcia Lambas D. 1987, *MNRAS*, **228**, 501.
Okumura S.K., Makino J., Ebisuzaki T., Ito T., Fukushige T., Sugimoto D., Hashimoto E., Tomida K., and Miyakawa N. 1992, in *Proceedings of the twenty-fifth Hawaii International Conference on System Sciences*, **I**, 151.
White S.D.M. 1990, in *Dynamics and Interactions of Galaxies*, ed. R. Wielen, p. 380, Springer-Verlag.

Group Simulations: Looking for Compact Groups

KIRK D. BORNE
STScI, Homewood Campus, Baltimore, MD 21218, USA

HAROLD F. LEVISON
Southwest Research Institute, San Antonio, TX 78238, USA

ABSTRACT: A fast simulation algorithm for loose groups is under development. We will use it to study the evolution of loose groups, the formation rate of compact groupings within such groups, the frequency of compact chance alignments, and the merger and interaction rates within these environments.

1. BASIC THEORIES OF COMPACT GROUPS

Compact groups of galaxies are among the densest environments in which we find galaxies. Because of their relatively low masses (a few galaxies, as compared to hundreds of galaxies in rich clusters), compact groups typically have low velocity dispersions. These two facts together (high density, low dispersion) argue for a very short time scale for the dynamical evolution of such groups. The end result of this rapid evolution is believed to be the merger of all of the bound galaxies into a single stellar system. That we see compact groups at all is therefore surprising from a dynamical point of view. There are a few ways that one can think about this unexpected observational fact: (1) The groups really exist (i.e. they may be bound or unbound dense groupings, or the cores of collapsing loose groups, or the cores of galaxy clusters, or ...). According to Mamon (these proceedings), this may account for perhaps $\sim$ 20% of the catalogued compact groups. (2) The groups don't exist as physical groupings of galaxies (i.e. they may be chance alignments, or H II regions, or ...). According to Mamon (these proceedings), this may account for $\sim$ 80% of the compact groups. ($2\frac{1}{2}$) They can't exist (i.e. interactions are too frequent, or mergers are too rapid, or ...). We can reduce these three alternative viewpoints to two basic questions:
(A) Do compact groups exist? ... which must be addressed by observation.
(B) Can compact groups exist? ... which must be addressed by theory. This is the focus of our work.

2. SIMULATIONS OF GROUPS AND SUBGROUPS

Our particular modeling approach is to take the techniques, wisdom, and insights developed from studies of binary galaxy simulations and to apply these to the study of slightly larger groupings of galaxies. We have performed a large number of binary galaxy simulations over the past few years and have finely tuned the techniques for studying such one-on-one encounters (Borne et al. 1988, Balcells et al. 1989, Borne 1990a,b, Borne and Richstone 1991). There are a couple of significant results (that are relevant here) that we and many other authors have identified from the study of such simulations: (1) Tidal effects can be very strong and very long-lived, even in fast encounters. (2) As galaxies merge, the tidal disturbances become more randomly distributed in phase, thus leading to a more symmetric appearance of the remnant (in both morphology and kinematics).

Thus, even in a group environment, binary encounters between galaxies and their subsequent dynamical evolution and remnant settling can usually be treated essentially independently from the rest of the galaxies in the group during the time of the encounter. The dynamical time of the binary system is often much shorter than the crossing time within the surrounding loose group, which further isolates the binary evolution from that of the group as a whole. We therefore can study the evolution of loose groups by focussing most of the computational energy on the binary encounters and by treating the rest of group with a much coarser computational algorithm. Close random encounters involving 3 or 4 galaxies can also be treated within this framework – as a rapidly evolving independent subgroup within the slowly evolving loose group environment.

We are now developing a numerical simulation code tailored to this type of study. It will basically be a two-level "tree" algorithm. We will treat each galaxy as a single particle when they are far apart, using large time steps, and labeling each galaxy particle by its internal properties (mass, spin, binding energy). Then, when two (or more) galaxies come close together (within a fixed number of galactic radii), each of the "galaxy particles" involved in the encounter will be replaced by a large-N galaxy model having the requisite internal properties. The close encounter will then be evolved (with smaller time steps) until completion (i.e. until either the galaxies merge or the galaxies separate on an unbound trajectory). In this way, we hope to run a very large number of group simulations and to test the various hypotheses related to the state and fate of loose groups.

3. TESTING THE COMPACT GROUP THEORIES WITH GROUP SIMULATIONS

The chance alignment hypothesis suggests that compact groups consist of binaries seen in projection with other group members (Mamon, these proceedings). The "reality" hypothesis suggests that the compact groups can somehow survive until the present epoch or are just now forming. Whatever the explanation, one must run a large ensemble of group simulations to test the hypotheses. Our plan is to carry out this task with a very fast hybrid code. This code will be a combination of (1) a small-N loose group simulation code (with $\sim$10-20 "galaxy particles") and (2) a large-N "binary" simulation code (with $\sim 10^4$ particles per galaxy). This is essentially our two-level tree code. The advantages of this method are both numerical (i.e. it is efficient) and physical (i.e. it employs a plausible treatment of both the close encounters and the slower group evolution).

Some of the issues that we hope to address from our study of a large ensemble of group simulations are these: Are the compact groups chance alignments between binaries and other members of looser groups? How strong and frequent are the binary encounters in loose groups? What are the interaction and merger rates within groups? Can compact groups evolve out of loose groupings by dynamical friction and/or multiple-body encounters? Can tidal features survive in dense environments? Can compact groups live a long time (with multiple encounters "breaking up the dance") before the merger instability wins?

4. FINAL REMARKS AND OPINIONS

We see few very bright isolated ellipticals (... the hypothesized end-products of compact group dynamical evolution). Maybe compact groups are just now forming, or perhaps they can really survive a lot longer than we think. Or maybe they simply cannot exist; the groups we see may be non-physical.

We see very few clear cases of pure binary interactions in compact groups (i.e. with strong, long-lived tidal features). Maybe compact groups are not chance alignments involving binaries. Of course, frequent encounters in truly dense environments will act to disrupt extended tidal tails and to phase-mix other tidal deformities.

Group simulations may support the theory that compact groups cannot exist: there are very few WYSIWYG's ("What You See Is What You Get"; i.e. compact groupings in 3-dimensions). Therefore, maybe only $\sim 20\%$ of the groups are real (Mamon, these proceedings). This may be necessary because true compact groups would merge too fast and because the rate of strong tidal encounters would be too high and therefore inconsistent with the interaction frequency seen in current group catalogs.

If compact groups cannot exist and are not otherwise real (cluster cores, collapsing loose groups, ...), then the compact groups must be chance alignments.

This sounds like circular reasoning, and it is all very inconclusive. In any case, both theoretical and observational tests are necessary to avoid applying the bias of WYWIWYG's ("What You Want Is What You Get"). Though we have no results yet to report, we hope that our simulations will ultimately provide some insight into this critical area of research on the nature, evolution, and fate of groups of galaxies.

5. REFERENCES

Borne, K.D., Balcells, M., and Hoessel, J.G. 1988, *ApJ*, **333**, 567.
Balcells, M., Borne, K.D., and Hoessel, J.G. 1989, *ApJ*, **336**, 655.
Borne, K.D. 1990a, in *Dynamics and Interactions of Galaxies*, ed. R. Wielen, Berlin, Springer, p. 196.
Borne, K.D. 1990b, in *Paired and Interacting Galaxies*, IAU Colloquium No. 124, eds. J.W. Sulentic, W.C. Keel, and C.M. Telesco, Washington, NASA, p. 537.
Borne, K.D., and Richstone, D.O. 1991, *ApJ*, **369**, 111.

A Search for new dwarf members of the M81 group in the 21cm line of HI

W.K. HUCHTMEIER
Max-Planck-Institut für Radioastronomie,
Auf dem Hügel 69, W-5300 Bonn 1, Germany

E.D. SKILLMAN
Astronomy Department, University of Minnesota,
116 Church St., SE, Minneapolis, MN 55455, USA

1. INTRODUCTION

The Las Campanas survey of the Virgo cluster (Binggeli et al. 1985) is about the only complete catalog down to low luminosities. The Local Group and the catalog of nearby galaxies (Kraan-Korteweg and Tammann 1979) are easily affected by incompleteness (e.g., Irwin et al. 1990), especially in the zone of avoidance. Studies of nearby groups of galaxies are usually confined to only a small part of the sky and, due to the groups' small distances, are not sensitivity limited.

Our interest in this project is twofold. One reason is to find very low mass galaxies for abundance studies. Without velocity information it is impossible to distinguish between truly low mass galaxies and low surface brightness background galaxies. Skillman et al. (1988, 1989a,b) have shown that extremely low mass galaxies hold the most promise for finding regions of near-primordial abundances. HI line-widths give an additional clue to discriminate against higher mass systems. Follow-up CCD H-alpha imaging of the most promising candidates will lead to the identification of HII regions for chemical abundance studies.

The second goal of the program is a study of the M81 group dynamics. HI spectra provide the only practical method for obtaining radial velocities of the very low surface brightness galaxies. A large population of dwarf galaxies with known radial velocities will allow a new estimate of the mass distribution of the M81 group. The finding of Bothun et al. (1987) that a previously cataloged dwarf galaxy in the Virgo cluster is indeed a massive low surface brightness background galaxy stresses the need for radial velocity measurements in such studies.

In the case of the M81 group, several catalogs contain information on dwarf galaxies; we have used the catalogs by Karashentseva (1968), Börngen and Karashentseva (1982) and Börngen et al. (1982) who list 39 faint probable members of the M81 group excluding the known bright galaxies that usually are taken to identify this group (e.g., de Vaucouleurs 1975). These dwarf galaxies have been further classified concerning probable membership by Karachentseva, Karachentseva, and Börngen (1985). Lo, Sargent, Kowal and Smith (1986) have recently completed a deep IIIa-J survey of the M81 group using the 48 inch Palomar Schmidt. 137 dwarf galaxies were detected of which 57 were judged to be group members. We composed a search list from the above named catalogs, which amounts to 158 dwarf galaxies roughly in the range $65° \leq Dec. \leq 80°$ and $7h30m \leq R.A. \leq 13h$. Most of these dwarfs have crude visual magnitudes of 17 to 19 mag (absolute magnitudes of −13 to −11 at the distance of the M81 group), diameters around 1 arc min or less, and classifications of type Im with a few E type or spheroidals (Börngen spheroidals were not detected in HI). A few galaxies are very close to brighter galaxies of this sample. This confusion could not be resolved as most probably the detections are due to the brighter galaxies.

2. NEW EFFELSBERG HI OBSERVATIONS

Observations were performed in 1987/88 and 1991 using the 100m radiotelescope at Effelsberg which has a half power beam width of 9 arc min for the 21 cm HI line. A FET receiver in the first run and a HEMT receiver in the latter run yielded a total system noise of 60K and 30K respectively. The total power mode was applied observing an on-source position followed by an empty reference field which was subtracted from the former in order to reduce instrumental effects. Pointing is believed to be about ±15 arc sec r.m.s. Some of the optical positions are not known to better than to 1 arc min. However, this is small compared to the telescope beam of 9 arc min. Galaxies with optical velocities were observed with a velocity range of 1250 km/s (with a resolution of 6 km/s). The radial velocity range of the M81 group overlaps with velocities of local hydrogen. This "confusion" essentially could not be solved. Fainter signals adjacent to the range of velocities of local hydrogen were checked by both frequency modulation and rudimentary mapping. All such features which were suspected to be due to galaxies were found to be extended considerably larger compared to known extended haloes of galaxies and therefore considered to be local emission.

We aimed for a detection limit of 10^6 solar masses in HI at the distance on M81. For comparison the dwarf M81dwA (Lo and Sargent 1979) has an HI flux of 4.2 Jy km/s (line widths of 22 km/s) which corresponds to $7.2\ 10^6$ solar masses.

In order to identify background galaxies, a larger range of radial velocities had to be searched. The 1024 channel autocorrelator was split into four channels using a bandwidth of 6.25 MHz each resulting in a resolution of 6 km/s and a velocity coverage from -400km/s to 4000 km/s. Galaxies not detected in this velocity range were searched at higher radial velocities.

3. SENSITIVITY

The typical rms noise was in the range 0.003 to 0.01 K; assuming line widths of 15 or 30 km/s (which are typical for small dwarf galaxies) and three times the rms noise this corresponds to an HI line integral of 0.1 to 0.6 Jy km/s or to 0.25 to $1.5\ 10^6$ solar masses of HI at the distance of the M81 group. This represents a detection limit much better than needed for the M81dwA system. However this galaxy remains at the lower end of the HI fluxes of the detected objects. 73 galaxies (a detection rate of 45%) were detected; most of which are background objects. Fig.1 shows the radial velocity distribution of these galaxies which is like that of a magnitude limited sample of galaxies rather than that of a group of galaxies at a certain distance.

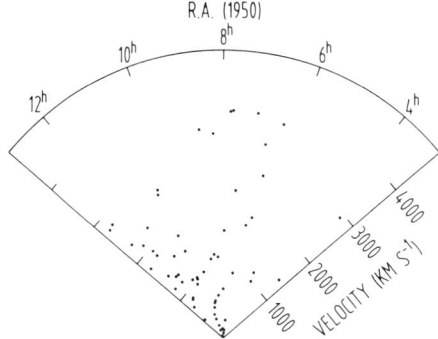

FIGURE 1. The detected galaxies are plotted as a function of right ascension and radial velocity in this cone diagram. Only the very center ($v \leq 300 km/s$) is the M81 group.

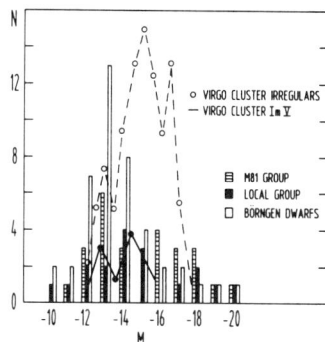

FIGURE 2. The luminosity function of different samples of galaxies. The Virgo Im V dwarfs (filled circles) lies in the same luminosity range as dwarfs from the Local group as well as the M81 group.

4. THE M81 GROUP

Collecting data on all galaxies that were named members or possible/probable members we compiled a list of 25 galaxies with HI observations (this work and the HI catalog, Huchtmeier and Richter 1989). In Fig.1 we present a cone diagram for the detected galaxies. Summed over the declination range 65 to 80 degrees the galaxies are displayed in R.A. versus radial velocity. Only the very center of this plot (v ≤ 300 km/s) represents the M81 group. This cone diagram does not show a complete sample of galaxies, hence this diagram does not necessarily document the structure of the nearby universe. It is evident from this figure that all detected galaxies are far more distant than the M81 group.

The rest of the analysis is restricted to galaxies which are true members of the M81 group. In Fig.2 we plot the luminosity function of different samples of dwarf galaxies. The Virgo cluster irregulars (Binggeli et al. 1985) are given by open circles, irregulars of luminosity class V (filled circles) are to be found in the range $-12 \leq M \leq -15$. This luminosity range is well frequented by the

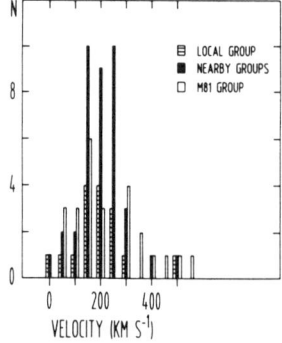

FIGURE 3. The velocity distribution of different groups is given. The velocity dispersion of the M81 group seems to be about twice as large as for the comparison groups. This fact seems to point as a contamination of the M81 group by background galaxies.

other samples in this figure:
  a) the Local group of galaxies (de Vaucouleurs 1975),
  b) the "standard" M81 group (galaxies with radial velocities),
  c) the M81 group (Börngen dwarfs).

Unfortunately we got only crude magnitudes for some galaxies in this sample. The magnitudes given for the M81 group (Börngem dwarfs) assume all of them at the distance of M81. However, some of them will be in the background and hence will be more luminous. The overall shape of the distribution of the local group and the M81 group is quite good. The comparison samples a) and b) peak in the same magnitude range as the Virgo Im V sample, a fact that suggests some agreement in the luminosity function of the dwarf galaxy population in clusters and groups. The velocity distribution for some nearby groups is given in Fig.3. The distributions of individual groups were centered to each other using an arbitrary zero point of the velocity axis. Those groups are
  a) the Local group of galaxies,
  b) the sum of four nearby groups of galaxies (Centaurus, CVnI, CVnII, and M101),
  c) the M81 group (allowing galaxies in the velocity range $-200 \leq v_{LG} \leq 400 kms^{-1}$).

The comparison samples a) and b) exhibit a typical velocity dispersion of the order of 100 km/s, whereas the corresponding values of the M81 group seems to be about twice as high. This seems to be evidence of contamination with background galaxies. Excluding three galaxies with radial velocities greater than 300 km/s reduces the velocity dispersion to about "normal" i.e., comparable to the other groups shown Fig. 3.

5. CONCLUSIONS

This survey of dwarf galaxies in the M81 group area revealed:
  1) a detection rate of 45%, most of which are background galaxies relative

to the M81 group, and

2) no new HI rich dwarfs in the M81 group. With the detection limit in mind, this result is unexpected. HI-observations of the Centaurus and Sculptor groups (Freeman et al. this conference) yield 15 to 25 new HI rich dwarfs around those groups with a detection limit of about $1\ 10^7$ solar masses.

While it is true that the velocity range covered by local neutral hydrogen prohibits detection of faint galaxies within this range, we would still expect at least a dozen or so dwarf galaxies to be detectable in the M81 group if it had a luminosity function similar to that of the Centaurus and Sculptor groups. The undetected galaxies might be either low surface brightness galaxies in the background or dwarf ellipticals in the M81 group itself.

In their study on dwarf galaxies Binggeli et al. (1990) conclude that "dwarf galaxies obey a morphology-density relation like the giants", and "dwarf irregulars and dE's prefer strongly dense environments on all scales". Hence the different dwarf population of the Centaurus and Sculptor groups on one side and the M81 group on the other side might just be documenting the large variety in the population of groups.

6. REFERENCES

Binggeli, B., Sandage, A., and Tammann, G.A. 1985, *AJ*, **90**, 1681.
Binggeli, B., Tarenghi, M., and Sandage, A. 1990, *A&A*, **228**, 42.
Börngen, F., and Karachentseva, V.E. 1982, *Astron.Nachr.*, **303**, 189.
Börngen, F., Karachentseva, V.E., Schmidt, R., Richter, G.M., and Thanert, W. 1982, *Astron.Nachr.*, **303**, 287.
Bothun, G.D., Impey, C.D., Malin, D.F., and Mould, J.R. 1987, *AJ*, **93**, 29.
deVaucouleurs, G. 1975 in *Galaxies and the Universe*, eds. A. Sandage, M. Sandage, and J. Kristian, University of Chicago Press, Chicago, pp. 557.
Huchtmeier, W.K., Richter, O.-G. 1989 *A General Catalog of HI Observations of Galaxies*, Springer-Verlag, New-York.
Irwin, M.J., Bunclark, P.S., Bridgeland, M.T., and McMahon, R.G. 1990, *MNRAS*, **244**, 16p.
Karachentseva, V.E. 1968, *Comm. Byurakan Obs.*, No. **39**, 61.
Karachentseva, V.E., Karachentseva, L.D., and Börngen, F. 1985, *A&AS*, **60**, 213.
Kraan-Korteweg, R.C., and Tammann, G.A. 1979, *Astron.Nachr.*, **300**, 181.
Lo, K.Y., and Sargent, W.L.W. 1979, *ApJ*, **227**, 756.
Lo, K.Y., Sargent, W.L.W., Kowal,C.K., and Smith, X. 1986, *BAAS*, **18**, 450.
Sandage,A., Binggeli,B., Tammann, G.A. 1985, *AJ*, **90**, 1759.
Skillman, E.D., Melnick, J., Terlevich, R., and Moles, M. 1988, *A&A*, **196**, 31.
Skillman, E.D., Terlevich, R., and Melnick, J. 1989a, *MNRAS*, **240**, 563.
Skillman, E.D., Kennicutt, R.C., and Hodge, P.W. 1989b, *ApJ*, **347**, 875.

Properties of Radio Groups

ESTHER L. ZIRBEL[1]
Department of Astronomy, Yale University, New Haven, CT 06511

ABSTRACT: In this paper we determine if radio galaxies are the first-ranked ellipticals in groups. Furthermore, we analyze if radio groups can be used to study the general evolution of galaxies in groups.

1. INTRODUCTION

We wish to study the evolution of galaxies in groups over a wide range of redshifts. Because on average radio galaxies inhabit denser regions than galaxies chosen at random (e.g., Longair and Seldner 1979, Seldner and Peebles 1978, Prestage and Peacock 1990), they may provide us with a suitable method of finding groups of galaxies in an unbiased fashion. Therefore, we chose a sample of radio galaxies, which is limited only by their radio luminosities and their redshifts. However, prior to studying the evolution of galaxies in our sample of radio groups, we need to know if our method introduces further biases. Therefore, it is crucial to obtain a better understanding of the radio phenomenon itself, and particularly of the type of groups that we select. The main purpose of this paper is to determine if radio groups are representative of groups in general. If this is not the case, we shall search for a link between the group environment and the radio phenomenon.

Radio galaxies are a special subset of galaxies, because by definition they have large amounts of radio emission and they are believed to have an active nucleus which produces non-thermal radiation. It is already clear that not every galaxy can become radio loud; only the brightest elliptical-like galaxies may turn into powerful radio sources. In the literature, we often encounter a generally accepted statement which we quote here: "Radio galaxies are the first ranked ellipticals in groups". This statement makes three different assumptions: (1) Radio galaxies are elliptical galaxies; (2) Radio galaxies are always in groups; (3) Radio galaxies correspond to the first ranked galaxies. In this paper, we shall address all 3 points separately.

2. SAMPLE DESCRIPTION

We tried to limit ourselves to a narrow range in radio powers, ranging from 10^{26} to 10^{28} Watts Hz^{-1} (measured at 408 Hz), and to two redshift intervals ranging from z=0.03 to z=0.22 and from z=0.3 to z=0.5. CCD photometry of fields containing the radio sources, and of nearby fields was obtained in the B and V bands for the low-redshift sample and in the V and R bands for the high-redshift sample. Altogether our sample consists of 96 fields. Whenever possible, we will

---

[1] PRESENT ADDRESS: SPACE TELESCOPE SCIENCE INSTITUTE, HOMEWOOD CAMPUS, BALTIMORE, MD 21218.

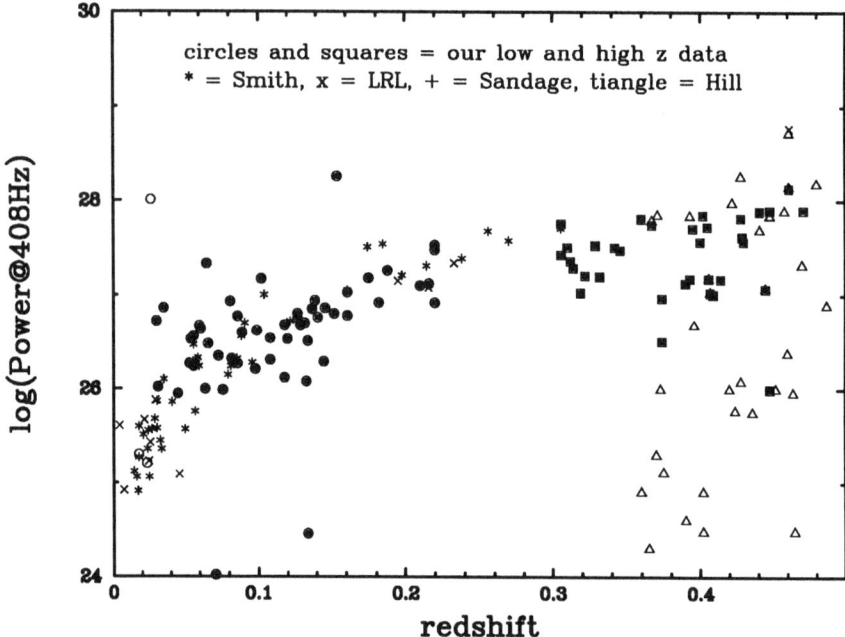

FIGURE 1. Power/redshift relationship for all galaxies in our sample.

supplement our high redshift data with sources taken from Hill (1990), and our low redshift data with sources from Longair, Riley, and Laing (1983), Smith (1989), and Sandage (1973).

In Figure 1 we display the redshift/radio power diagram for all sources, where filled symbols correspond to our own data, while open symbols are a combination of Hill (1990), Longair, Riley, and Laing (1983, LRL), Smith (1989), and Sandage (1973). The data are incomplete in the area below the mean $P - z$ relation, because the samples (taken from the 3CR, 4C, 5C, Bologna, 1–Jansky, and Parkes catalogs) are flux-limited. Since the radio luminosity function is steep, any sample will be dominated by the intrinsically least luminous sources which are above the selection limit. In any flux-limited sample, the luminosity of objects above the flux limit increases rapidly with redshift, producing a strong $L - z$ correlation. Furthermore, the sample appears incomplete in the area above the mean $P - z$ relationship, due to the volume effect. Hence, in the rest of this analysis, we have to remember that we have an artificial $P - z$ relationship in our data, which will make interpretations difficult.

The radio galaxies in our sample are all extended double-lobe sources that have a variety of radio morphologies. The radio emission may be stronger in one lobe than in the other, or, it may be stronger either at the far ends of the lobes or closer to the center. Fanaroff and Riley (1974, FR) first attempted to classify their structure and divided the sources into two classes (FRI and FRII) based on the position of the hotspots relative to the total extent of the source. The FR class is quantitatively defined as the ratio between the distance

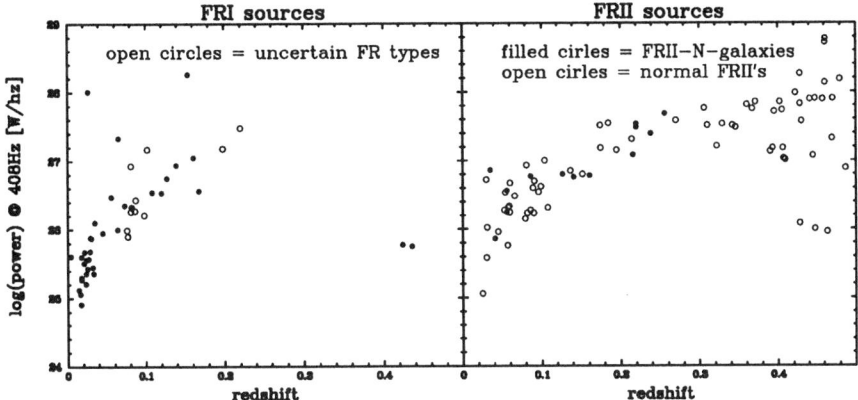

FIGURE 2. Power/redshift relationship for FRI and FRII sources.

between the highest brightnesses on opposite lobes to the total extent of the sources, as measured from the lowest radio contour lines. For FRI's this ratio is less than 0.5; for FRII's it is larger than 0.5. The intention of this system is to discriminate between "classical" double lobe sources in which the maxima in the brightness distribution are located towards the leading edge of the source components (e.g., Cygnus A), and "complex" sources which generally have a much more diffuse radio structure sometimes without hot-spots. We obtain the radio morphologies for our sources from the literature.

In Figure 2 we plot once more the radio power–redshift relationship, but with radio morphology identifications. We see that the frequency of FRI galaxies decreases with radio power, while that of FRII's increases. This is not new and was already noted by Fanaroff and Riley (1974). However, we also see that the frequency of FRI's decreases with redshift. In other words it is not clear if epoch, or radio power, or both, or something else determine the radio structure of the sources. Nevertheless, it is interesting to note that there are no high-redshift powerful FRI's. At higher redshifts only two radio sources (5C12.71 and 5C12.168) were positively identified as FRI's by Hill and Lilly (1990). However, their radio powers are relatively low. In fact, at lower redshifts some FRI's are even more powerful than the most powerful FRII's at the same redshift. Therefore, we propose that the conditions for producing powerful FRI's are more favorable at low redshifts.

Here, we need to mention that the gas characteristics may play a crucial role in the radio phenomenon. However, this will not be discussed here. On one hand the gas may be responsible for supplying the fuel for the radio source, and on the other hand it may determine the outlook of the radio lobes. In powerful sources, the jet may be supersonic, which may give rise to edge-brightened (FRII) morphologies (e.g. Norman et al. 1984); if the jets are subsonic, the radio emission will be closer to the central object, with diffuse emission gradually disappearing with distance from those regions, and an FRI source may be observed. While the jet power is crucial, so is the gas density. Therefore, any changes in the gas density of the environment may thus give rise to changes in the radio morphologies.

3. ANALYSIS

3.1. Is the radio galaxy always an elliptical galaxy?

Traditionally, powerful radio sources (with powers above 10^{24} Watts Hz^{-1}) were thought to be associated with bright elliptical galaxies. However, based on large photometric surveys of radio ellipticals by Lilly, McLean, and Longair (1984), Heckman et al. (1985), Baum et al. (1988), Smith (1988), Djorgovski et al. (1985), and Djorgovski (1985), it has already become clear that they often have rather unusual structures, extended envelopes, or very bright nuclei. The most well studied sources (Fornax A, Cygnus A, and Centaurus A) have spectacular optical morphologies and also dust features; some other radio galaxies have clear stellar disks (Sansom et al. 1987) and others may be severely distorted (Heckman et al. 1986), possibly due to galaxy interactions. Smith (1988) claims that the morphological disturbances are often quite large (several kpc) in extent and varied in their form, taking shapes of tidal tails, fan, shells, and asymmetric isophotes. In order to understand the importance and the frequency of these optical peculiarities, we produce surface brightness profiles of radio galaxies. Since normal elliptical galaxies may also have some structure, it is most important to compare these surface brightness profiles to a sample of normal elliptical galaxies. A good comparison sample of normal elliptical galaxies is provided by Schombert (1984). He showed that ellipticals can be described by a family of generalized profiles, where the structure of the galaxy is a function of its luminosity. These generalized profiles give us the opportunity to compare the structure of our radio galaxies both quantitatively and qualitatively to their radio-quiet counterparts.

From the surface brightness profiles of our radio galaxies, we see that they can have a variety of different shapes. Therefore, we shall classify these profiles. The most obvious categories are galaxies which look roughly elliptical, or others which have cD features, or N-type morphologies. In the table below, we list these categories.

We see that one quarter of the radio galaxies are N-galaxies, another quarter are either cD's and double nuclei galaxies, while the remaining galaxies have mostly elliptical-like structures. However, more than half of those ellipticals have relatively disturbed surface brightness profiles. In summary, less than a quarter of all radio galaxies have truly elliptical surface brightness profiles.

Dividing our sample according to FR morphologies, we see that almost all N galaxies are FRII's and that they constitute at least one third of the FRII sample. Half of the remaining FRII sources are normal ellipticals, while the others have relatively disturbed profiles. On the other hand, the FRI sources are found in CD-like or double-nuclei systems. The general trend is clear: FRI sources prefer systems that have been exposed to a large degree of galactic processing.

3.2. Is the radio galaxy always in a group?

This test is easy to perform. We merely count all galaxies that surround the radio galaxy within an annulus of 0.5 Mpc and that are brighter than -19th magnitude. We shall refer to this quantity as the "richness" of the group, and

TABLE 1. Surface brightness profile classifications

	Total #	% of all	FRI(#)	FRII(#)
All sources	72	100.0	16	29
Ellipticals	12	17.1	3	7
E+S	18	25.7	2	7
N–galaxy	15	21.4	1?	11
fE	6	8.6	2	2
cD	11	15.7	5	1
Double	5	7.4	4	0
Contaminated	3	4.3	0	0
Misclassified	2	2.7	0	2

denote it by $N_{0.5}^{-19}$. We also include Hill's (1988) high-redshift groups, whose values have been transformed to our scale. The procedure of calculating $N_{0.5}^{-19}$ has been described in detail by Allington-Smith et al. (1993; AEZO).

In Figure 3, we display the histogram for the distributions of group richnesses. The radio groups span richnesses ranging from negative values (due to background oversubtraction) up to $N_{0.5}^{-19} = 70$ for the richest group. In Figure 3 we see that many, but not all radio galaxies, live in groups; 25% of our radio galaxies have 2 members or less. The main point is clear: the majority of radio galaxies do live in groups; however, a non-negligible amount of radio galaxies prefer the empty field.

While we claim that most radio galaxies live in groups, the relevant question to ask is: how likely are they to be found in a group as compared to non-radio-active galaxies? This question was first addressed by Seldner and Peebles (1979) who cross-correlated the 3C source counts with Lick counts. They found that in general, radio galaxies are found in density-enhanced regions. This statement is often interpreted to mean that radio galaxies have a higher frequency of neighbors than a random population of normal galaxies, which need not be the case. In AEZO, we approached this issue differently. We compared the distribution in group richness of radio groups to groups from the CFA survey (Geller and Huchra 1983). We found that the distribution in group richness for radio-loud and radio-quiet groups are comparable for groups that are richer than $N_{0.5}^{-19} \approx 12$, but that poorer radio groups are relatively underabundant.

Next, we shall analyze this behavior for the FRI and FRII groups. Since we are not interested in the space distribution of the radio groups, we shall normalize the distributions at a richness of $N_{0.5}^{-19} = 12$. The results are displayed in Figure 4. We see clearly that, as compared to the mean distribution of radio groups, the FRI sources preferentially live in richer environments, and that they are found even less frequently in poor groups than the average generic radio galaxy. Low redshift FRII groups are different: they avoid the richest groups

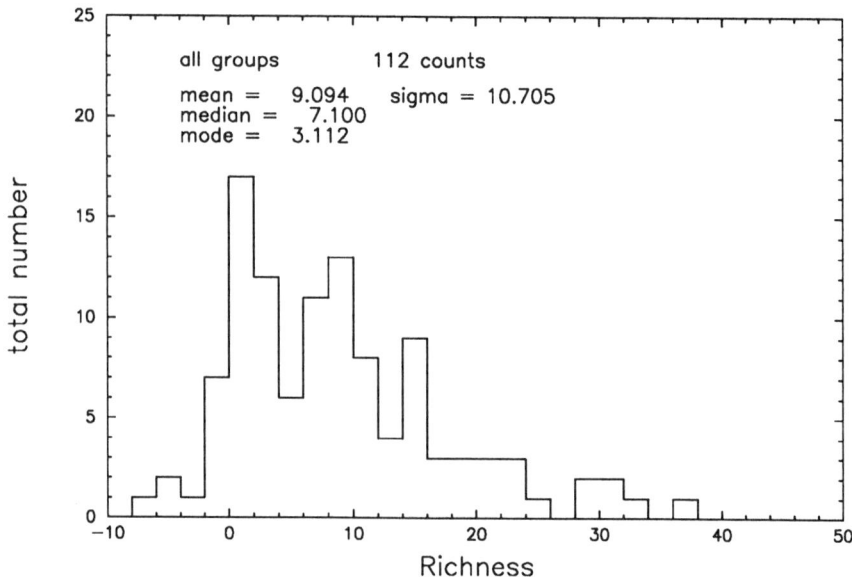

FIGURE 3. Richness distribution of our radio groups, including Hill's sample.

altogether, but poor FRII groups are still not as abundant as normal groups. Now we are able to clarify some contradictions in the literature between Heckman et al. (1985), Hill (1989), Prestage and Peacock (1989,1990), and Longair and Seldner (1979); namely, that many FRI's do inhabit poor environments, but that in average a larger fraction of FRII's are in even poorer environments.

Next, we shall study the evolution of the environment of the radio groups, but only for FRII groups. There are significant differences between the high- and the low-redshift samples. A KS-test (Figure 5) shows that the two distributions are drawn from the same parent population only at the 6.75% level. The mean richnesses of high and low redshift FRII groups are 11.4±2.1 and 5.8±1.2, respectively. While low-redshift FRII groups avoid rich environments totally, they do exist at higher redshifts. In AEZO, we explained that this result is qualitatively in general agreement with Hill and Lilly (1990).

However, we have to be cautious with possible interpretations, because of the artificial epoch dependence of radio power. If we subdivide our sample into high and low powers, we find that weak FRII's also avoid rich groups. So we have to determine which is more fundamental: the fact that low-redshift FRII's avoid rich groups, or that low-power FRII's avoid rich groups. A KS-test shows that the high- and low-power samples belong to the same parent distribution at the 49.98% confidence level. Since the KS-probability is lower for the high- and low-redshift samples (6.8%), we conclude that the epoch dependence is more fundamental than the power dependence, and we confirm that low-redshift FRII's avoid rich groups.

We know that the richnesses of groups cannot decrease as radically as seen within our FRII sample. Indeed, if merging of galaxies occurs, the mean richness

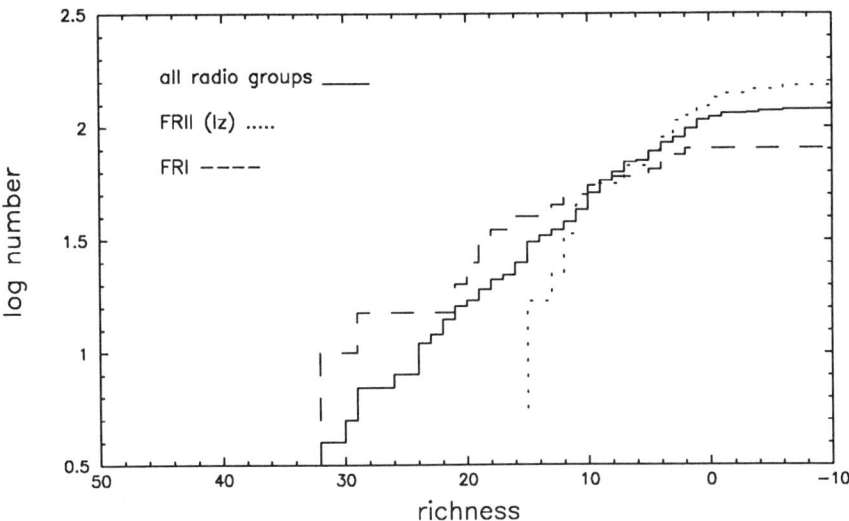

FIGURE 4. Cumulative richness distributions for FRI and FRII groups.

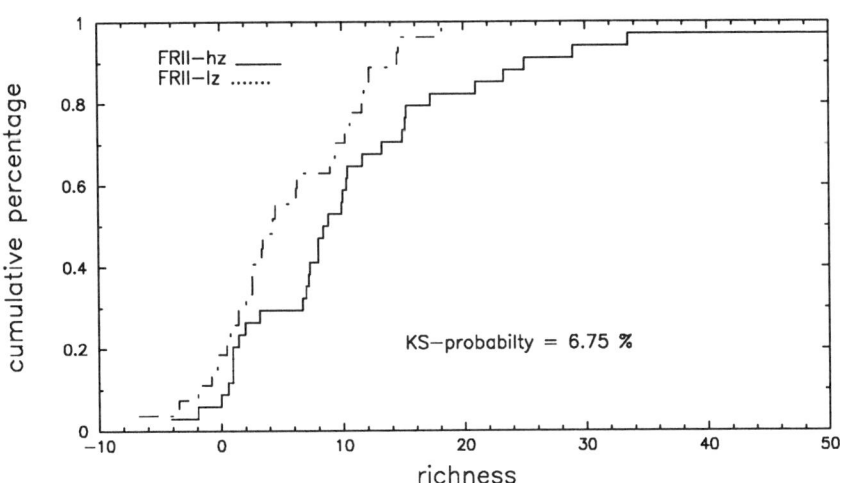

FIGURE 5. KS-test of the richness of high and low redshift FRII groups.

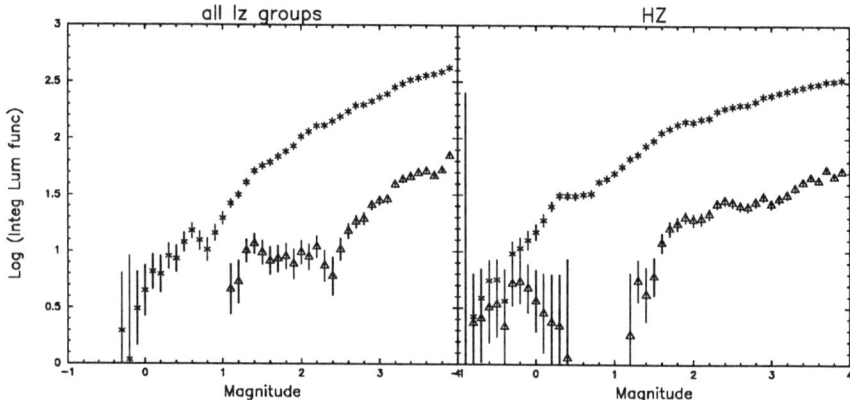

FIGURE 6. Normalized, background-subtracted luminosity function for the high and low redshift groups (stars — all galaxies, triangles — only blue galaxies).

may decrease, but only slightly. It cannot halve itself in one third of a Hubble time. In fact groups are exposed to galactic infall, which increases the group richness. Therefore, the above result implies that the radio phenomenon itself has evolved. If the FRII phenomenon is relatively long-lived, it suggests that sources in rich environments have terminated their evolution. On the other hand, if the FRII phenomenon is short-lived, it implies that the environmental conditions for producing FRII sources in rich groups have changed in the last third of a Hubble time.

In summary, we conclude that poor radio groups are less abundant than poor normal groups. Furthermore, this tendency is even stronger for FRI sources and is epoch-dependent for FRII's.

3.3. Is the radio galaxy the first-ranked galaxy in a group?

It has always been assumed that the radio galaxy is the brightest group member, however no formal tests have been performed. If truly only the brightest member can turn into a radio galaxy, it means that there is something very special about this galaxy, and also that the environment is critical in the formation of the radio source.

With our sample we can test easily if the radio galaxy is indeed the first-ranked galaxy in its group. Since our CCD frames are contaminated by foreground galaxies, we have to devise a statistical method to determine if they are group members or not. We use the background-subtracted luminosity function, which we normalize to the brightness of the radio galaxy. In Figure 6 we display the resulting luminosity functions for the high- and low-redshift samples separately. Among the 74 low-redshift groups we see that there are 3 galaxies that are brighter than the radio galaxy, while among the 34 high-redshift groups this number has increased to 12.

The 3 bright galaxies in the low-redshift sample belong to 3C225a, 3C424, and PKS1214. 3C225a is by far the optically faintest galaxy (-20.68), and it has a rather unusual surface brightness profile. We suspect that it is a misidentification. The 3C424 groups has 2 brighter galaxies. One galaxy is at a separation of only 30 kpc, and we suspect that 3C424 is a dumbbell galaxy, while the other brighter galaxy may be a foreground object. The last low-redshift group is PKS1214, with another brighter galaxy that is at a projected distance of 0.48 Mpc. We identify that galaxy as another radio source, namely PKS1215. Their redshifts are comparable (0.0775 and 0.0765) and on the CCD frames it appears that the two groups are merging. Note that there is some controversy in the literature about the fluxes measured for PKS1214 and PKS1215 (e.g., Schilizzi and McAdams 1975; Burbidge 1967; Clark, Bolton, and Schimmers 1966). It appears uncertain if the double-peaked radio signal should be associated with one or both galaxies.

For the high-redshift sample we produce the luminosity function, but only for the blue galaxies. We define a galaxy to have a blue color if it is 0.2 magnitudes bluer than an elliptical galaxy of the same absolute magnitude. We see that many of the galaxies that are more luminous than the radio galaxies have blue colors. If starbursts are responsible for producing a bluer color, this means that the galaxy would be fainter otherwise. Nevertheless, there are still 5 galaxies in the high-redshift sample that are brighter than the radio galaxy. Of these, 3 belong to 4C29.44, 1 to 4C27.51, and 1 to 5C6. Clearly, for these galaxies we would need redshifts.

In summary, within our low-redshift sample, there is only one group (3C424) that may have a brighter member. The identification of 3C225a needs clarification, and PKS1214 may not be a radio galaxy. Therefore, we conclude that among the 74 low-redshift groups, in 71 cases (96%) the radio galaxy is a confirmed brightest group member. Note that this is a lower limit. Within the high-redshift sample the corresponding numbers are 31 of 34 groups (91%).

3.4. The link between environment and the radio phenomenon.

In Figure 7 we plot the total radio outputs for FRI and FRII sources against their group richnesses. Only for the FRI's, we observe a clear trend between richness and radio power, where stronger sources are embedded in richer groups. This correlation is significant at the 99.97% level. However, many optically more luminous FRI's are also more powerful (Owen and White 1991), and the optically more luminous FRI's preferentially live in richer environments (Zirbel 1992). Therefore, one may argue that FRI's in richer environments appear to be more powerful only because they are more massive, and the more massive ones are also more powerful. However the relationship between magnitude and richness is much weaker (at the 92.5% level) than the relationship between radio power and richness. This suggests that the FRI sources are more powerful, because they reside in richer groups. For FRI's, we have thus established an intimate relationship between the environment and the radio phenomenon. Furthermore, we suggest that analyzing the gas contents of groups may be important to understanding the FRI phenomenon.

For FRII's, we do not observe any correlations between the radio power, the magnitudes, the galaxy sizes, or the group richness. Therefore, we suggest that the formation of FRII's may be a random process, perhaps triggered by stochastic galaxy encounters. We test this hypothesis by separating our sample into galaxies that have disturbed surface brightness profiles and/or a close

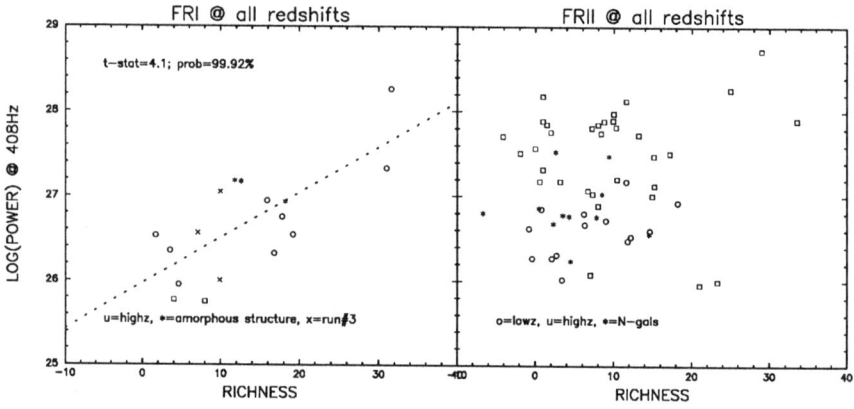

FIGURE 7.   Power/richness relationships.

neighbor and into isolated galaxies. In Figure 8, we plot the power distributions for these samples, and we find that disturbed radio galaxies are more powerful by a factor of 2.4±1.8. This implies that galaxy interactions are important and that they enhance the total radio output.

4. SUMMARY

1) We confirm that radio galaxies are mostly found in groups. While 25% have two group members or less, poor radio groups are found less abundantly than normal groups. This result may have interesting implications for the radio phenomenon: Either the probability of making a radio galaxy in richer groups is more favorable, or the lifespan of the radio activity is shorter in poorer regions which gives us the impression that we observe fewer. Nevertheless, radio galaxies do provide us with a good method to find groups of galaxies at higher redshifts.
2) In at least 95% of all radio groups, the radio galaxy is indeed the brightest group member. Many of the higher redshift galaxies that are brighter than the radio galaxy, may be undergoing a violent starburst, which makes them momentarily brighter.
3) Radio galaxies are at best elliptical-like galaxies. Only about 17% are true ellipticals, another 21% are N-galaxies, 26% have strongly disturbed surface brightness profiles, and another 33% are processed galaxies (eg cD's, double nuclei, or fE's).
4) Since there are no strong FRI sources at earlier epochs, and because the number of FRI sources decreases with redshift, we will claim that they formed recently. They are preferentially found in cD-like or double-nuclei systems that have already undergone a large degree of galactic processing, perhaps even prior to becoming a radio galaxy. Furthermore, we found that FRI galaxies in richer environments are larger, more luminous, and also more powerful. This implies that FRI sources form in richer groups and, furthermore, it suggests that the total radio power output of FRI's is pre-determined by the richness of the group

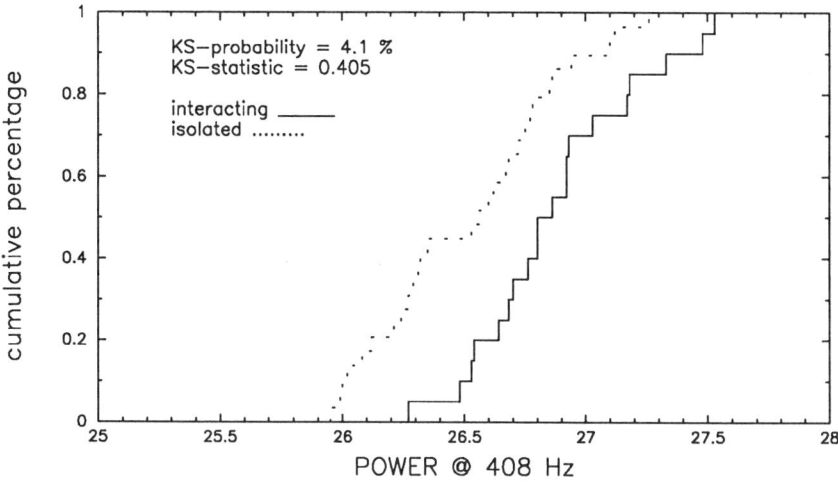

FIGURE 8. Radio power distributions for interacting and isolated radio galaxies.

in which it forms.

5) For FRII's we do not find significant correlations between the richness of the environments and the radio power, the absolute magnitude, or the optical size of the underlying galaxy. However, we do find that low-redshift, and/or low-power FRII sources avoid rich groups. We propose two explanations for this behavior: (a) The FRII's are long-lived and in rich groups — they have terminated their evolution either because the evolution in richer groups is faster (perhaps due to quicker gas exhaustion), or because they formed at earlier epochs. (b) If FRII's are short-lived, then the environmental conditions for triggering the FRII phenomenon may have changed only in rich groups. In both cases powerful FRII's will eventually die, leaving behind a radio-quiet elliptical. If the FRII phenomenon is indeed short-lived, the lack of any significant correlations between radio power, magnitudes, size, and group richness can only be explained if the formation of FRII's is a random process, perhaps triggered by stochastic galaxy encounters. Indeed, we found that interacting FRII sources are in average more powerful than isolated systems.

6) And finally — Can we use radio galaxies as a selection criterion to study the evolution of galaxies in groups? We have shown that FRI and FRII groups are different subsets of groups in general. The underlying galaxies have different properties, they live at different epochs and in different environments. Therefore, we can only use radio groups to study the evolution of groups if we know how to describe the characteristics of these two subsets. In fact, we may even be able to use these characteristics to constrain certain evolutionary theories. For example, at higher redshift, FRII groups inhabit a larger range in richnesses. While this result may be correlated to the radio phenomenon itself, it may also tell us something significant about group evolution in general. For instance, it may imply that environmental conditions in rich groups have changed during the last third of a Hubble time, and/or it may suggest that the evolution in

richer groups is quicker.

5. REFERENCES

Allington-Smith, Ellis, Zirbel, & Oemler 1993, Ap.J., 404, 521 (AEZO)
Baum S.A., Heckman T., Bridle A., van Breugel W., Miley G., 1988, Ap.J.Suppl., 68, 643
Burbidge, 1977, Ap.J.Lett., 149, L51
Clark, Bolton, Schimmers, 1966, Austral.J.Phys, 19, 375
Fanaroff B.L., Riley J.M. 1974, MNRAS, 167, 31P
Geller and Huchra 1983, Ap. J. Supp., 52, 61
Heckman T.M., Carty T.J., Bothun G.D., 1985, Ap.J., 288, 122
Hill G. 1989 PhD Thesis University of Hawaii
Hill G., Lilly S.J. 1991, Ap.J., 376, 1
Hine, Longair M.S. 1979, MNRAS, 188, 111
Laing, Riley, Longair M.S. 1983, MNRAS, 204, 151
Lilly S.J., Mclean I.S., Longair M.S. 1984, MNRAS, 209, 401
Longair, M.S. 1966, MNRAS, 133, 421
Longair M.S., Seldner M. 1979, MNRAS, 189, 433
Norman M.L., Winkler K.A., Smarr L. 1984, In Proc. NRAO Workshop No. 9S, p150
Owen F.N., Laing R.A. 1989, MNRAS, 238, 357
Owen F.N., White R.A. 1991, MNRAS, 249, 164
Prestage R.M., Peacock J.A. 1989, MNRAS, 230, 131
Sandage A. Ap.J. 1972: 178,1; 178,25; 1973: 183,711; 183,743
Schombert, J. 1984, PhD thesis, Yale University
Schilizzi, McAdams 1975, Mem RAS 79, 1
Seldner, M., Peebles, P.J.E. 1978, Ap.J. 225, 7
Smith E., 1988, PhD Thesis, University of Maryland
Zirbel E.L., 1992, PhD Thesis, Yale University

Workshop Summary

GARY A. MAMON
DAEC, Observatoire de Meudon, 92195 Meudon, FRANCE

1. 1. INTRODUCTION

Groups of galaxies, despite a variety of defining algorithms, contain nearly half of the galaxies in the Universe, more than any other subsystems (*e.g.,* de Vaucouleurs 1975). This realization has sparked interest in these systems as tracers of the mean mass density of the Universe (Gott and Turner 1977). The consequent quest for the typical group mass-to-light ratio has gained impetus with the advent of group catalogs defined from 3D galaxy surveys, as pioneered by Huchra and Geller (1982, hereafter HG) and Geller and Huchra (1983, hereafter GH).

Among the groups that appear the most compact in projection on the plane of the sky, the 100 compact groups of Hickson (1982, hereafter HCGs) appear to be at least as dense as the cores of rich clusters of galaxies. HCGs may thus be the densest isolated systems of galaxies in the Universe, and hence the ideal place to look for galaxy interactions.

This workshop is the first devoted exclusively to groups of galaxies, and the points made above, as well as the numerous studies, both observational and theoretical, presented at this meeting, justify the importance of groups of galaxies as tools for studying both galaxy dynamics and cosmology.

In the first part of this *summary* (§2), I will review what was said at this workshop, and then, in §3, I will comment in some detail on some issues related to the topics covered in §2.

2. WHAT WAS SAID

2.1. Loose groups

Besides weighing the Universe, Richter pointed out that groups serve as distance indicators used in the interpretation of the large-scale velocity-field (Lynden-Bell *et al.* 1988; Bertschinger *et al.* 1990), and moreover represent an important link in the hierarchy of large-scale structures between binary galaxies and both (poor) clusters and what Tully (1987) calls clouds. And Richter also remarked that loose groups have the property of not emitting much X-ray radiation, although the relatively compact 'loose' group no. 39a in Turner and Gott's (1977) catalog does show diffuse X-ray emission (Biermann *et al.* 1982).

Studies of groups are hampered by their sensitivity to the chosen definition and selection algorithm used. Turner and Gott (1977) provided the first objective group defining algorithm, based on surface density enhancements in the Zwicky catalog. Huchra and Geller (1982) pioneered the *friends-of-friends* (hereafter, FoF) approach, in which a galaxy lying close (in redshift space) to a previously defined group is a member of this group. In an alternative approach called *dendogram*, pioneered by Materne (1978), groups are built by linking two *close* entities, where these can be galaxies or groups built in a lower hierarchy.

Here, close is defined on the two entities (*e.g.*, Tully 1987) or on the full catalog (Materne 1978). In both FoF and dendograms, one can build a hierarchy of systems, depending on some selection parameters such as the mean 3D group density in the FoF algorithm. The main difficulty lies in deciding what *close* means in angular-redshift space.

A serious problem with the CfA group FoF algorithm is that it is adapted to conserve the group multiplicity function as a function of flux. So groups near the magnitude limit of the parent galaxy catalog are allowed to have larger galaxy separations, both in projected physical space, and in redshift space. The virial theorem then tells us that the group M/Ls are a function of magnitude. A promising way to circumvent this problem has been introduced by Nolthenius and White (1987), who searched for CfA like groups in N-body simulations of pieces of Cold Dark Matter universes. Moore presented a refinement of this method (Moore, Frenk and White 1992). The projection effects can then be minimized by fine tuning the parameters in the selection algorithm to optimize the appearance of the selected groups in *real*-space wedge diagrams (where clusters are roundish concentrations rather than fingers of God). Unfortunately, as pointed out by Moore, contamination can not be brought below 20%. It would be interesting to see this work repeated with very different group defining algorithms, such as used by Tully (1987) to see which group finding algorithm produces more realistic groups.

In any event, group catalogs are small (typically < 100 groups with 4 or more galaxies), and it is important to build new ones. This is why the group catalog presented by Tucker, based upon a galaxy survey made at Las Campanas (with Kirschner, Oemler, Schechter and Schectman) is a welcome addition to the loose group database. Although it goes deeper than previous group catalogs, it is very different in one respect: it has a *bright-end* magnitude cutoff. The bright galaxies will be easily added to the survey in the near future.

A comparison of three group catalogs was presented by Perea (with Moles and coworkers in Granada). The group mass-density is inferred from the size, taken as $(R_{-1}R_2^2)^{1/3}$, where R_i is the ith moment in the distribution of galaxy projected separations. Then the loose groups of GH, HG and the Hickson's (1982) compact groups (hereafter HCGs) represent a sequence of groups with increasing densities and decreasing mass-to-light ratios.

3. COMPACT GROUPS

Although Shakhbazyan compact groups of compact galaxies are selected for having mostly galaxies that are compact on the red POSS plates but fuzzy on the blue plates, Moles showed that the galaxies in these groups are giant elliptical galaxies, with normal colors, with few dwarfs. Moreover, Moles computed the mass-to-light ratio to be $M/L \simeq 40\,h$ for both the groups (just as Hickson 1990 computed for HCGs) and the individual galaxies, and found group crossing times of 3% of H_0^{-1}.

Six talks at this workshop were devoted to observations of Hickson compact groups. Analysis of CCD images shows that the HCG ellipticals have the usual $r^{1/4}$ law surface magnitude profiles (Bettoni; Perea), normal colors, except for a few cases of blue ellipticals (Zepf; Bettoni; Moles), presumably sites of recent bursts of star formation, but which are almost always *low-luminosity* ellipticals, which is not what you expect for merger remnants. This fraction of blue ellipticals is smaller than for isolated pairs of galaxies (Moles). Zepf reported that although many HCG ellipticals have irregular isophotes, few have *boxy* isophotes thought to be signs of recent merging (Binney and Petrou 1985).

Bettoni reported that the giant ellipticals in HCGs were slow rotators as elsewhere (Illingworth 1977), but added that the HCG giant ellipticals follow an $L \sim \sigma_v^{6.2}$ relation instead of the usual $L \sim \sigma_v^4$ relation (Faber and Jackson 1976). Moreover, Zepf reported that giant ellipticals have velocity dispersions 20% lower than in clusters, whereas Djorgovski, Weir and de Carvalho (1992) report velocity dispersions up to 5 times lower than for normal ellipticals of the same luminosity.

The optical luminosity function of HCGs was discussed by Sulentic, who, in contrast with a study by Mendes de Oliveira and Hickson (1991), chose not to extrapolate to the parent distribution of HCGs, once selection effects are considered, but instead used the philosophy of *what you see is what there is*. Thus he finds that the faint-end slope of the galaxy luminosity function is -1 as in the field (Efstathiou *et al.* 1988; Loveday *et al.* 1992), instead of -0.3 found by Mendes de Oliveira and Hickson, but he adds that there is a real lack of dwarf galaxies. This lack of dwarfs is probably caused by the magnitude selection criterion rejecting faint galaxies.

Williams presented her analysis of the distribution and kinematics of neutral gas in HCGs, and reported a variety of situations: In one case, an HCG is probably a single galaxy, made up of four HII regions. In the other cases, HI is distributed around individual galaxies, two or three galaxies, or the whole group, which Williams suggests represents a sequence of increasingly evolved groups.

A summary of the observations of HCGs was given by Mamon. In his subsequent analysis of the theories competing to explain the nature of compact groups, Mamon pointed out that chance alignments of galaxies within loose groups (and clusters) are at least ten times more frequent than bound 3D compact and isolated subgroups, and that over 70% of these 1D chance alignments contain binaries or triplets, thus explaining quite accurately the $\simeq 50\%$ of HCG galaxies that show signs of dynamical interaction. So he claims that most compact groups are not bound dense quartets.

4. GALAXY MORPHOLOGIES

Whitmore reported a sharp rise of the fraction of ellipticals towards the center of clusters (Whitmore and Gilmore 1991), argued that this rise results from the suppression of disk formation in the cluster cores by *tidal* processes. He further suggested that the hot intergalactic gas emitting in X-rays corresponds to the baryonic material that failed to infall and dissipate into disks.

Charlton presented a study of morphologies in pairs of galaxies, and reported that in dense environments, ellipticals tended to have companions. Although her work is in collaboration with Whitmore, Charlton argued for a merger scenario to model the evolution of the Hubble sequence, where the ratio of galaxy masses determines the morphology of the merger remnant. In Charlton's model, mergers with ratios up to 3:1 give ellipticals, and between 3:1 and 20:1 you get an S0 (this last ratio is in rough agreement with Tóth and Ostriker's (1992) 4% mass accretion limit that a spiral galaxy can sustain and still preserve its thin disk.

There was alot of talk about dwarf galaxies. Indeed, while Freeman and Huchtmeier reported on successful observational searches for dwarfs in loose groups, dwarfs seems to lack in Shakhbazyan (Moles) and Hickson (Sulentic) compact groups.

Zirbel reviewed the properties of radio galaxies in groups, and found that radio-galaxies are preferentially situated in rich environments. ¿From the distribution of the radio-emission, she listed eight sources for the nature of these radio-galaxies, from normal ellipticals to cDs to double dumbshell objects. When radio sources are associated with HCG ellipticals, these are almost always the brightest group member (Menon and Hickson 1985; Menon 1990).

5. DYNAMICS

Groups are a good place to study dynamical interactions such as merging. When a cannibal gobbles up its victim, its merger cross-section increases, thus increasing the probability of merging with another galaxy. Menci presented a model (Cavaliere et al. 1991) that follows the effects of this merging instability 1981) in groups on the evolution of the group mass function, and reported a runaway evolution. In reality, one must also consider tidal processes, which prevent cannibals from becoming too large, and thus slows down significantly this merging instability (Mamon 1987). Outside of groups, the evolution of the mass function can also be studied, in the context of the Press-Schechter (1974) approach, and Menci reported that in such a modelling, dwarfs form early and the low-end power-law slope of the mass function reaches -1.2 instead of Press and Schechter's -2. This result seems strange to this summarizer as the Press-Schechter approach supposedly incorporates naturally the effects of mergers, and as such the slope should remain at -2, and the shallower slope (-1 to -1.3) of the *luminosity* function comes from the details of the infall of gas into the potentials of dark halos (see Blanchard, Valls-Gabaud and Mamon 1992).

The kinematics of substructures in rich clusters and their effect on galaxy morphologies were addressed by Fitchett and Whitmore, respectively. As Fitchett pointed out, substructures are basically groups (albeit very rich ones), and a group's orbit will decay faster into the cluster core, by dynamical friction, then for individual galaxies. In some cases, such as Hydra, clusters may really be two clusters superposed, and the substructure is really a projection effect. In such as case, the cluster velocity dispersion that is measured on the full cluster will be an overestimate. This could be good news for the cold dark matter theory of galaxy formation, which has difficulty in making high velocity dispersion clusters, such as Abell 2319, which Fitchett reported to have $\sigma_v = 1819 \, \text{km s}^{-1}$. Finally, the observation of *speeding cDs* found to move rapidly within their cluster may be caused by merging in a rapidly moving subcluster.

Nevertheless, substructure in clusters seems to have little importance on determining galaxy morphologies. Indeed, as mentioned by Whitmore, the observed morphology-density relation is unaltered when one performs an angular scrambling of the galaxy positions (Sanromà and Salvador-Solé 1990), and and the local fraction of ellipticals in clusters seems at least as dependent on clustocentric radius than on local density (Whitmore and Gilmore 1991).

Athanassoula presented preliminary trends in new N-body simulations of groups using a massively parallel machine in Japan called GRAPE, putting 50 000 particles in the group. Despite the long time required to follow a group (1 day of CPU), she managed to realize thirteen runs! With a dark matter to luminous matter ratio of 3/1, she finds that when the dark matter is in individual halos, the merging of galaxies occurs *slower* than when the dark matter is in a common intergalactic envelope. This trend is opposite from that in previous work: Mamon (1987) found that in dense groups, individual halos overlap and merge rapidly, whereas when the dark matter is in a common envelope, instead of direct merging, one has to wait for the galaxy orbits to decay by dynamical friction towards the group center where they eventually merge. The effect found by Athanassoula may reflect the small extent of her halos.

Another interesting approach for group simulations was presented by Borne, who is building a code that will be self-consistent only when two galaxies are very close, but will reduce the effect of each galaxy to a point in its center otherwise. This can be interpreted as an extreme form of a tree code (Barnes and Hut 1986; Hernquist 1987), where there are only two levels of interaction (close and far). This type of code is worth testing as it promises to provide the first semi-self-consistent *statistical* set of group simulations. Indeed, large simulated databases are necessary to explore parameter space, estimate the statistical distribution of evolution times, and search for very rare events, and so far only one such work has been published (Mamon 1987), but it was not self-consistent, in the sense that there was only *one* particle per galaxy (and all the missed physics was *explicitly* added in).

In an attempt to model the properties of galaxies with satellite companions, Zaritsky has modelled the halos of spirals with 1D simulations of top-hat spherical infall (White and Zaritsky 1992). These simulations confirm the detailed semi-analytical models of Fillmore and Goldreich (1984) and Bertschinger (1985). He compares the distribution of simulated and observed (Zaritsky *et al.* 1992) radial velocity differences, and finds that halos must extend to at least 200 kpc around large spiral galaxies, and that the angular momentum at maximum expansion is less than 90% of the maximum possible value (for circular orbits).

Cosmological simulations of pieces of the Universe were presented by Moore (with Frenk and White) and Gelb (with Bertschinger), both using particle-particle-particle-mesh (P^3M) codes with 64^3 and up to 256^3 particles, respectively (*i.e.*, 250 000 and 20 million). Gelb points out that if all interactions are dissipationless (*i.e.*, there is no gas) then there is an *overmerging* problem: groups merge into clusters (*Godzillas* for those who favor Japanese animated literature) much faster than is inferred from the observed distribution of circular velocities. With gas properly added in (which is technically very difficult) the overmerging disappears. This is confirmed in simulations by Evrard, Summers and Davis (1992). But does this mean that all previous simulation work without gas, such as that presented by Moore, is to be thrown away?

6. REMARKS

6.1. How can one want to work on loose groups?

It takes plenty of courage to work on loose groups for the following reasons:

1) the membership, properties (density, M/L) and dynamical state of groups strongly depend on the defining algorithm used.

2) In most catalogs, a substantial fraction of loose groups are not (yet) virialized (Byrd and Valtonen 1985; Giuricin et al. 1988). In each catalog, there is a mix of dynamical states: the different groups are expanding, collapsing, virialized, or coalescing (see Giuricin et al. 1988).

3) Loose groups are probably not well isolated from their surroundings, and one must therefore simulate large pieces of the Universe to simulate well the properties of groups.

A promising approach is to tie together problems of group membership and groups as distance indicators, by defining groups in such a way that the scatter correlation plots for distance indicators is minimized (Weigelt and Kates 1990). Another interesting avenue is the use of velocity fields to determine the dynamical state of groups (Byrd and Valtonen 1985; Kiseleva and Orlov 1993).

6.2. Compact groups are becoming so popular!

Since the publication in 1982 of Hickson's catalog of compact groups, the number of studies of these apparent dense systems has increased dramatically, so that today, compact groups are perhaps the best studied systems of galaxies. Of course, I like to think that the surge in the number of publications on the topic dates from my own article where I suggested, against the current general belief, that HCGs were mostly chance alignments within loose groups (Mamon 1986). In reality, the surge in the number of observational studies comes from the belief that HCGs contain the *Holy Grail*: extraordinary dynamics. HCGs turned out to be indeed interesting, but not quite as expected.

To understand the nature of compact groups *one must compare compact group properties to those of loose groups, pairs, and isolated galaxies*. This brings the following plea: *Members of Observing Time Allocation Committees: Please understand that the observations of control samples are crucial in understanding the nature of compact groups!*

Although I presented hopefully convincing arguments on why most HCGs are caused by chance alignments within loose groups and clusters, one can ask what are the main weaknesses with this chance alignment hypothesis. I will answer that 1) the density of HCG environments will require careful modelling; and 2) one must understand the abnormal $L - \sigma_v$ relations reported by Bettoni and Zepf.

Finally, there is much hope in analyzing digitized galaxy catalogs to search for compact groups, as done by Iovino and collaborators (1991). Indeed, such studies have the advantages that: 1) they allow objective selection; 2) they incorporate a well-defined limiting magnitude; 3) they permit to establish how compact group properties vary as a function of varying the selection criteria. As such one will know whether HCGs are a special class of objects, as advocated by Sulentic, or whether they represent a particular case in a continuous family of groups.

6.3. Are binary galaxies chance alignments within loose groups?

After pushing the idea that compact groups are chance alignments within loose groups, let me ask "are pairs of galaxies caused by chance alignments in loose groups?" Nearly half of the binaries in the catalogs of Turner (1976) and Schweizer (1987) belong to groups or clusters. How probable is it that a binary within a loose group is real? To answer this, Brieu and Mamon (1993) have simulated groups with the group code published in Mamon (1987), looked, in projection, for binaries that met Turner's and/or Schweizer's isolation criteria. It turns out that 40% to 80% of these binaries were *unbound*, depending on whether the dark matter in the group was in individual halos or in a common envelope. Moreover, the unbound binaries occupy roughly the same region in diagrams of projected separation vs. radial velocity difference than the bound pairs. This makes hopeless any attempt to analyze the mass-distribution and extent of galaxy halos using binaries within groups. In other words, the isolation criteria used by Turner and by Schweizer are insufficient to isolate binaries from the groups in which they may reside. We have also recovered the distribution of eccentricities of the real binaries, necessary for detailed dynamical analyses.

6.4. N-body simulators are finally waking up!

There are many more studies involving cosmological simulations of large-scale structure in the Universe than there are studies in which the dynamics of groups are is simulated. But from the reports at this workshop, it seems that groups may finally receive the attention they deserve from the simulation community. Although this field is dominated by the study by Barnes (1989) and more recently Governato, Bhatia and Chincarini (1991), we learned at this workshop that Athanassoula, Moore (with Frenk and White) and Gelb (with Bertschinger) have jumped into the group simulation band-wagon, with additional pledges by Borne (with Levison), Zepf (with Frenk), and Fitchett.

An outstanding issue, raised at this meeting to understand compact groups is the question: *how long do features of dynamical interaction, such as tidal tails and shells survive in a dense group?* Simulations of virialized dense groups should provide some answers soon. Cosmological simulations have the advantage of generating initial conditions for groups linked to well-defined theoretical principles. But if Gelb is correct (and I fear that he is) one must add gas in these simulations, which makes this type of study extremely difficult! Moreover if one want to study *compact* groups, one needs unparalleled length resolution, and as mentioned above, decent statistics too. In summary, I believe that we are nowhere near to have new simulations solve everything, but we will not progress if we do not try!

6.5. What makes ellipticals in clusters?

Finally, one must admit that Whitmore's theory for the enhanced fraction of ellipticals in the inner cores of clusters is very attractive. The proto-disk (and halo) of a galaxy will be stripped after it falls through a cluster, by the tidal field of the cluster potential (see White and Rees 1978). We know that the halos of $10^{12} h^{-1} M_\odot$ galaxies *should* extend to $700 h^{-1}$ kpc (that is the current turnaround radius) today (see Mamon 1992). These halos are severely tidally stripped after the first passage through the cluster. The baryonic material that collapses and forms the disk may come from the same size region, or may come from a smaller one (White 1991). But if disks form early around galaxies, before they fall into clusters, they should survive the tidal shock as galaxies passes through the cluster core. But the galaxies that are now near the cluster center are likely to have fallen in at an early time (see Mamon 1992), and therefore would not have had enough time to accrete significant disks. Hence the trend reported by Whitmore.

One can think of two alternatives. First, *if* galaxy mergers cannot directly produce the sharp rise in elliptical fraction in the inner cores of clusters, there is a way they can explain it indirectly, and here is where compact groups and ellipticals in clusters are tied in. If the primordial density fluctuations follow Gaussian statistics, then at early epochs, dense groups will naturally form near the centers of proto-clusters. Then these groups will coalesce by rapid merging into huge ellipticals, perhaps cDs. Once the cluster is formed, the remnants of the primordial dense groups will lie near their centers. This argument is exactly the same as the one used by Evrard, Silk and Szalay (1990) who estimated what morphology-density relation would come from considering elliptical galaxies as arising from high-density peaks of the primordial density fluctuations, but here I've replaced *ellipticals* by *dense groups*.

The second alternative is more straightforward, One can estimate the rate of mergers in clusters that grow by cosmological infall. Because the velocity dispersions of clusters are large, only a small fraction of the galaxy collisions will lead to mergers. But I've shown that this rate of mergers is not only far from negligible, but I can reproduce both shape and normalization of the morphology-density relation of Postman and Geller (1984) from a simple analytical integration of the merger rates. Soon after this workshop, I computed the morphology-*radius* relation and, to my surprise, it fit (Mamon 1992) Whitmore's (1993) relation very well, the fraction of ellipticals in clusters rising as $R^{-9/8}$.

Which of these three models is the correct one? One must (again) wait for detailed dynamical simulations, which for Whitmore's will require a detailed treatment of the dissipational collapse of the gas into disks. But perhaps cDs are the results of *multiple* mergers in dense groups and the clusters with centrally located cDs might then be those clusters near the center of which formed a primordial dense group.

7. REFERENCES

Barnes, J.E. 1989, *Nature*, **338**, 123.
Barnes, J.E., and Hut, P. 1986, *Nature*, **324**, 446.
Bertschinger, E. 1985, *ApJS*, **58**, 39.
Bertschinger, E., Dekel, A., Faber, S.M., Dressler, A., and Burstein, D. 1990, *ApJ*, **364**, 370.
Blanchard, A., Valls-Gabaud, D., and Mamon, G.A. 1992, *A&A*, **264**, 365.
Brieu, P., and Mamon, G.A. 1993, in preparation.
Byrd, G.G., and Valtonen, M.J. 1985, *ApJ*, **289**, 535.
Cavaliere, A., Colafrancesco, S., and Menci, N. 1991, *ApJ*, **376**, L37.
deVaucouleurs, G. 1975, in *Galaxies and the Universe*, Stars and Stellar Systems, Vol. **IX**, eds. A. Sandage, M. Sandage and J. Kristian, Chicago, Univ. of Chicago Press, p. 557.
Djorgovski, S., Weir, N., and de Carvalho, R.R. 1992, in *Cosmology and Large Scale*

Structure in the Universe, ASP Conference Series 24, ed. R.R. de Carvalho, San Fransisco, A.S.P., p. 141.
Efstathiou, G., Ellis, R.S., and Peterson, B.A. 1988, *MNRAS*, **232**, 431.
Evrard, A.E., Silk, J and Szalay, A.S. 1990, *ApJ*, **365**, 13.
Evrard, A.E., Summers, F.J., and Davis, M. 1992, *Nature*, submitted.
Faber, S.M., and Jackson, R.E. 1976, *ApJ*, **204**, 668.
Fillmore, J.A., and Goldreich, P. 1984, *ApJ*, **281**, 1.
Geller, M.J., and Huchra, J.P. 1983, *ApJS*, **52**, 61.
Giuricin, G., Gondolo, P., Mardirossian, F., Mezzetti, M., and Ramella, M. 1988, *A&A*, **199**, 85.
Gott, J.R., and Turner, E.L. 1977, *ApJ*, **213**, 309.
Governato, F., Bhatia, R., and Chincarini, G. 1991, *ApJ*, **371**, L15.
Hernquist, L. 1987, *ApJS*, **70**, 419.
Hickson, P. 1982, *ApJ*, **255**, 382.
Hickson, P. 1990, in *Paired and Interacting Galaxies*, IAU Colloquium 124, eds. J.W. Sulentic and W.C. Keel, Washington, NASA, p. 77.
Huchra, J.P., and Geller, M.J. 1982, *ApJ*, **257**, 423.
Illingworth, G. 1977, *ApJ*, **218**, L43.
Iovino 1991, private communication.
Kiseleva, L., and Orlov, V. 1993, *MNRAS*, **260**, 475.
Loveday, J., Peterson, B.A., Efstathiou, G., and Maddox, S.J. 1992, *ApJ*, **390**, 338.
Lynden-Bell, D., Faber, S.M., Burstein, D., Davies, R.L., Dressler, A., Terlevich, R.J. and Wegner, G. 1988, *ApJ*, **326**, 19.
Mamon, G.A. 1986, *ApJ*, **307**, 426.
Mamon, G.A. 1987, *ApJ*, **321**, 622.
Mamon, G.A. 1992, *ApJ*, submitted.
Materne, J. 1978, *A&A*, **63**, 401.
Mendes de Oliveira, C., and Hickson, P. 1991, *ApJ*, **380**, 30.
Menon, T.K. 1990, *Dynamics and Interactions of Galaxies*, ed. R. Wielen, Berlin, Springer, p. 423.
Menon, T.K., and Hickson, P. 1985, *ApJ*, **296**, 60.
Moore, B., Frenk, C. and White, S.D.M. 1993, *MNRAS*, **261**, 827.
Nolthenius, R., and White, S.D.M. 1987, *MNRAS*, **225**, 505.
Postman, M., and Geller, M.J. 1984, *ApJ*, **281**, 95.
Press, W.H., and Schechter, P 1974, *ApJ*, **187**, 425.
Sanromà, M., and Salvador-Solé, E. 1990, *ApJ*, **360**, 16.
Schweizer, L. 1987, *ApJS*, **64**, 427.
Tóth, G., and Ostriker, J.P. 1992, *ApJ*, **389**, 5.
Turner, E.L. 1976, *ApJ*, **208**, 20.
Turner, E.L., and Gott, J.R. 1977, *ApJS*, **32**, 409.
Weigelt, U.V., and Kates, R.E. 1990, *A&A*, **240**, 1.
White, S.D.M. 1991, in *Dynamics of Galaxies and their Molecular Cloud Distributions*, IAU Symp. 146, eds. F. Combes and F. Casoli, Dordrecht, Kluwer, p. 383.
White, S.D.M., and Rees, M.J. 1978, *MNRAS*, **183**, 341.
White, S.D.M., and Zaritsky, D. 1992, *ApJ*, in press.
Whitmore, B.C. 1993, in *Physics of Nearby Galaxies: Nature or Nurture?*, XIIth Moriond astrophysics meeting, eds. T.X.T. Thuan, C. Balkowski and J. Trân Thanh Vân, Gif-sur-Yvette, Eds. Frontières.
Whitmore, B.C., and Gilmore, D. 1991, *ApJ*, **367**, 64.
Zaritsky, D., Smith, R., Frenk, C., and White, S.D.M. 1992, *ApJ*, submitted.